W9-ACU-966

Progress in Mathematical Programming

Nimrod Megiddo
Editor

Progress in Mathematical Programming

Interior-Point and Related Methods

Springer-Verlag
New York Berlin Heidelberg
London Paris Tokyo

Nimrod Megiddo
IBM Research Division
Almaden Research Center
San Jose, CA 95120-6099, USA

This work relates to Department of Navy Grant N00014-87-G-0037 issued by the Office of Naval Research. The United States Government has royalty-free license throughout the world in all copyrightable material contained herein.

Library of Congress Cataloging-in-Publication Data
Progress in mathematical programming / edited by Nimrod Megiddo.
p. cm.
"The starting point of this volume was a conference entitled 'Progress in Mathematical Programming,' held at the Asilomar Conference Center in Pacific Grove, California, March 1–4, 1987"—Forword.
Includes bibliographies.
ISBN 0-387-96847-4
1. Programming (Mathematics)—Congresses. 2. Linear programming—Congresses. I. Megiddo, Nimrod.
QA402.5.P785 1988
519.7—dc19 88-24788

Typeset by Asco Trade Typesetting Ltd., Hong Kong.
Printed and bound by R.R. Donnelley and Sons, Harrisonburg, Virginia.
Printed in the United States of America.

9 8 7 6 5 4 3 2 1

ISBN 0-387-96847-4 Springer-Verlag New York Berlin Heidelberg
ISBN 3-540-96847-4 Springer-Verlag Berlin Heidelberg New York

Foreword

The starting point of this volume was a conference entitled "Progress in Mathematical Programming," held at the Asilomar Conference Center in Pacific Grove, California, March 1–4, 1987. The main topic of the conference was developments in the theory and practice of linear programming since Karmarkar's algorithm. There were thirty presentations and approximately fifty people attended. Presentations included new algorithms, new analyses of algorithms, reports on computational experience, and some other topics related to the practice of mathematical programming.

Interestingly, most of the progress reported at the conference was on the theoretical side. Several new polynomial algorithms for linear programming were presented (Barnes-Chopra-Jensen, Goldfarb-Mehrotra, Gonzaga, Kojima-Mizuno-Yoshise, Renegar, Todd, Vaidya, and Ye). Other algorithms presented were by Betke-Gritzmann, Blum, Gill-Murray-Saunders-Wright, Nazareth, Vial, and Zikan-Cottle. Efforts in the theoretical analysis of algorithms were also reported (Anstreicher, Bayer-Lagarias, Imai, Lagarias, Megiddo-Shub, Lagarias, Smale, and Vanderbei). Computational experiences were reported by Lustig, Tomlin, Todd, Tone, Ye, and Zikan-Cottle. Of special interest, although not in the main direction discussed at the conference, was the report by Rinaldi on the practical solution of some large traveling salesman problems. At the time of the conference, it was still not clear whether the new algorithms developed since Karmarkar's algorithm would replace the simplex method in practice. Alan Hoffman presented results on conditions under which linear programming problems can be solved by greedy algorithms. In other presentations, Fourer-Gay-Kernighan presented a programming language (AMPL) for mathematical programming, David Gay presented graphic illustrations of the performance of Karmarkar's algorithm, and James Ho discussed possible embedding of linear programming in commonly used spreadsheets.

At the time of this writing, the area of new algorithms for linear programming is moving very fast. Interested readers are advised to check journals such as *Mathematics of Operations Research* and *Mathematical Programming*, and the bulletins of the joint national meetings of ORSA and TIMS, which continue to publish the most current information on the subject.

A report on the conference containing abstracts of the talks appeared as IBM Research Report RJ 5923 (59228) 10/29/87, and copies can be obtained from: IBM Thomas J. Watson Research Center, Distribution Services, Post Office Box 218, Yorktown Heights, New York, 10598, USA. IBM was also the host organization for the conference, and financial support was provided by the Office of Naval Research and by the IBM Corporation.

San Jose, California Nimrod Megiddo

Contents

Contributors

Clovis C. Gonzaga
COPPE Federal University of Rio de Janeiro, Cx Postal 68511,
21941 Rio de Janeiro, RJ, Brazil

Masakazu Kojima
Department of Information Sciences, Tokyo Institute of Technology,
Oh-Okayama, Meguro-ku, Tokyo 152, Japan

Nimrod Megiddo
IBM Research Division, Almaden Research Center, 650 Harry Road,
San Jose, CA 95120-6099, USA
and
School of Mathematical Sciences, Tel Aviv University, Tel Aviv, Israel

Shinji Mizuno
Department of Industrial Engineering and Management,
Tokyo Institute of Technology, Oh-Okayama, Meguro-ku,
Tokyo 152, Japan

J. L. Nazareth
Department of Pure and Applied Mathematics, Washington State
University, Pullman, WA 99164-2930, USA

J. A. Tomlin
IBM Research Division, Almaden Research Center, 650 Harry Road,
San Jose, CA 95120-6099, USA

Pravin M. Vaidya
AT&T Bell Laboratories, Murray Hill, NJ 07974, USA

Jean-Philippe Vial
Départment d'Économie Commerciale et Industrielle, Université de Genève, 2, Rue de Candolle, CH-1211 Genève 4, Switzerland

Yinyu Ye
Department of Management Sciences, University of Iowa, Iowa City, IA 52242, USA

Akiko Yoshise
Department of Industrial Engineering and Management, Tokyo Institute of Technology, Oh-Okayama, Meguro-ku, Tokyo 152, Japan

CHAPTER 1

An Algorithm for Solving Linear Programming Problems in $O(n^3L)$ Operations

Clovis C. Gonzaga

Abstract. This chapter describes a short-step penalty function algorithm that solves linear programming problems in no more than $O(n^{0.5}L)$ iterations. The total number of arithmetic operations is bounded by $O(n^3L)$, carried on with the same precision as that in Karmarkar's algorithm. Each iteration updates a penalty multiplier and solves a Newton-Raphson iteration on the traditional logarithmic barrier function using approximated Hessian matrices. The resulting sequence follows the path of optimal solutions for the penalized functions as in a predictor-corrector homotopy algorithm.

§1. Introduction

The complexity of linear programming problems was lowered to $O(n^{3.5}L)$ arithmetic operations by Karmarkar [8], beginning a new cycle in optimization research. The algorithm is based on solving an equivalent problem that has as objective his "logarithmic potential function," which was immediately recognized as a penalized objective function, his method being a smooth barrier function algorithm.

Each iteration of Karmarkar's algorithm is a steepest descent search with scaling, as was first noted in [13] and carefully formalized in [6]. This brought linear programming algorithms into the realm of nonlinear programming

This work was done while the author was at the Department of Electrical Engineering and Computer Science, University of California, Berkeley, CA 94720, USA.

Research partly sponsored by CNPq-Brazilian National Council for Scientific and Technological Development, by National Science Foundation grant ECS-8121149, Office of Naval Research contract N00014-83-K-0602, AFOSR grant 83-0361, the State of California Microelectronics Innovation and Computer Research Opportunities Program, and General Electric.

techniques, where smooth penalty function methods were then out of fashion because of the ill-conditioning that often happens near an optimal solution.

A second stage in this line of research led to the study of trajectories followed by several algorithms. The set of optimizers of the penalized functions using the well-known logarithmic barrier function with varying penalty multipliers, known as the "center trajectory," was described by Megiddo [10] and by Bayer and Lagarias [1]. This trajectory was recognized as an important region of the feasible set.

A breakthrough was obtained by Renegar [12], who used an approach based on methods of centers to follow the center trajectory. He achieved for the first time a speed of convergence leading to a solution of the problem in $O(n^{0.5}L)$ iterations, but each iteration needs $O(n^3 L)$ computations, with a total figure of $O(n^{3.5}L)$ computations, the same as in Karmarkar's algorithm. The same approach was subsequently followed by Vaidya [14], who proved a complexity bound equivalent to ours, obtained simultaneously and independently.

Our result is also based on following the center trajectory, but using a barrier function approach. The final result is a practical algorithm for which we expect good computational characteristics. The algorithm follows the trajectory by a short-step penalty function scheme and has the following advantages over the existing methods: there is no need to know the value of an optimal solution, and no lower bounds to an optimal solution are used during the application of the algorithm. No special format for the problem is needed, with the exception of the first iteration, which needs a point near the center. No projective transformations are used. Finally, the algorithm is well adapted to use large steps instead of the small steps needed for the convergence proofs.

Our algorithm follows the path of optimizers for penalized problems using the logarithmic barrier function. The same results could be obtained by using Karmarkar's potential function or the multiplicative potential: the curves would then be parametrized by the values of the lower bounds v in the expression

$$f_v(x) = n \log(c'x - v) - \sum_{i=1}^{n} \log x_i.$$

We chose the barrier function because the mathematical treatment is simpler, and so are the resulting procedures.

Barrier Function Methods and Homotopies

Logarithmic barrier function methods were introduced in [3] and systematically formalized in [2]. A small-step barrier function method tries to follow the path of optimizers for the penalized functions very closely, always staying comfortably in the region of convergence for the Newton method. This characterizes the method as a path-following procedure in the homotopy approach. Garcia and Zangwill have an extremely clear presentation of homotopies and path-following algorithms in [4]. The homotopy approach to barrier functions is also old, and the idea of following the path of optimizers was already present in [2]. An important reference for this study is the book by Lootsma [9], and recent results for linear programming problems were published by Megiddo [10]. Specifically, the barrier method is a predictor-corrector algorithm with the simplest of all schemes: each iteration is composed of an *elevator* step (decrease in the penalty parameter) followed by a Newton correction (internal minimization).

Although the homotopy approach provides nice interpretations of the behavior of the algorithm, we shall not profit much from it. This is because in complexity studies all *epsilons* and *deltas* must be carefully bounded, in opposition to what is done when deriving differential properties of the paths. We shall then follow a barrier function approach, which provides a natural setting for these precise computations.

The logarithmic barrier function approach with large steps was applied to linear programming in [5].

Instead of trying to adapt known results to our needs, we decided to write a self-contained paper and prove all results, except for the complexity of the projection operation needed at each iteration, proved by Karmarkar [8].

1.1. The Problem

We shall use a very general format for the linear programming problem, working on a bounded feasible set in the first orthant. One additional assumption must be made for the purpose of finding an initial penalty multiplier, and the easiest way to achieve this is to impose the presence of a simplex constraint. Adding a simplex constraint to a given problem is easy, as we comment later. We shall also study scaled problems that have the same format as the original one but without the simplex constraint.

Consider the following format for all linear programming problems to be studied here:

$$\begin{aligned} &\text{minimize } c'x \\ &\text{subject to } Ax = b, \\ &\qquad\qquad\quad x \geq 0, \end{aligned} \tag{1.1}$$

where $c, x \in \mathbb{R}^n$ and A is an $m \times n$ matrix, $m < n$. We assume that the feasible set is bounded.

Our algorithm will always work in the relative interior of the feasible set, and the following notation will be used:

$$\begin{aligned} S &= \{x \in \mathbb{R}^n \mid Ax = b, x > 0\}, \\ Q &= \{x \in \mathbb{R}^n \mid Ax = b\}, \\ D &= \{x \in \mathbb{R}^n \mid Ax = 0\}, \\ \mathbb{R}^n_+ &= \{x \in \mathbb{R}^n \mid x > 0\}. \end{aligned} \tag{1.2}$$

S is the relative interior of the feasible set, Q is its affine hull, $D = \text{Null}(A)$ is the set of feasible directions from any point in S, and $\mathbb{R}^n_+$ will denote the interior of the first orthant.

The Problem

The problem to be solved here is (1.1), with two additional assumptions: (i) that the simplex constraint is used, that is, for any feasible x, $e'x = n$; (ii) $e = [1\ 1\ \ldots\ 1]'$ is a nonoptimal feasible solution. If an initial feasible point is known, then these conditions can easily be obtained for an arbitrary problem (1.1) by means of a scaling, the introduction of a constraint of the form $e'x \leq M$, where M is a large number and two new variables. The complete procedure is detailed in Todd and Burrell [13].

Role of the Simplex Constraint

It is important to point out that the simplex constraint will be used at only one point, in Lemma 1.15, to choose the initial penalty multiplier. With the exception of the first iteration of the algorithm, the simplex constraint is not used at all, and nothing like Karmarkar's projective transformation is needed.

Projection Matrices

We shall denote by P the projection matrix onto D. The computation of this matrix for the scaled problems that we shall use is studied in [8].

1.2. Outline of the Algorithm

A barrier method breaks the constrained problem (1.1) into a sequence of nonlinear programming problems with the format

$$(P_k) \qquad \min_{x \in S} c'x - \varepsilon_k \sum_{i=1}^{n} \log x_i, \tag{1.3}$$

where ε_k are positive *penalty multipliers* such that $\varepsilon_k \to 0$. Problems (P_k) are constrained only by linear equality constraints and can thus be solved by algorithms for unconstrained minimization.

Following the methodology in Polak [11], we begin by studying properties of a *conceptual* algorithm and then evolve into an *implementable* method, followed by a *practical* algorithm.

A conceptual barrier function algorithm assumes that an exact solution x^k can be found for each (P_k). In this case it is well known that any accumulation point of the sequence (x^k) is an optimal solution for (1.1). The speed of convergence is dictated by the speed with which the multipliers converge to zero.

An implementable algorithm does not assume exact solutions. Each penalized problem is approximately solved and results in a point z^k that must be near the unknown x^k. This vector z^k will be the starting point for (P_{k+1}). If subsequent points x^k are near each other (in some sense to be explained later), then one Newton-Raphson iteration provides the good approximations that we need.

We begin by studying the conceptual algorithm and show that "short steps" are obtained by using a *fixed* sequence $\varepsilon_k = (1 - \sigma)^k \varepsilon_0$, where ε_0 is an initial penalty multiplier and $\sigma = 0.005/\sqrt{n}$. At this point it is already possible to prove the bound of $O(n^{0.5}L)$ for the number of iterations. This will be done in Section 3.

The next step, to be done in Section 4, is to show that good approximations to the conceptual solutions are obtained by scaling each penalized problem (P_k) about z^k and solving one Newton-Raphson iteration to generate z^{k+1}. A further improvement in complexity is obtained by using approximated Hessian matrices in these computations. A low-complexity procedure for achieving this is presented, as well as a method for computing the initial penalty multiplier.

Before we start studying the conceptual algorithm, Section 2 will be dedicated to listing some basic results on barrier functions and scalings.

All action will take place in $\mathbb{R}^n$. We shall denote vectors by lowercase letters, matrices by uppercase letters. The transpose of a matrix A will be denoted by A'. The vector with components x_i will be denoted by $[x_i]$, and the letter e will be reserved for the unit vector $e = [1\ 1\ \ldots\ 1]'$. For each vector z^k, an uppercase Z_k will denote the diagonal matrix $\operatorname{diag}(z_1^k, z_2^k, \ldots, z_n^k)$, to be used in scaling operations. The norms 1, 2, 3, and ∞ for $\mathbb{R}^n$ will be denoted by $\|\cdot\|_1$, $\|\cdot\|$, $\|\cdot\|_3$, and $\|\cdot\|_\infty$, respectively, and other norms will be defined in Section 2.

§2. Scalings, Barrier Functions, Quadratic Approximations

This section concentrates basic results related to logarithmic barrier functions and their quadratic approximations. They are mostly "common knowledge" results, and our purpose is to present the notation and specialize the results to our needs.

2.1. Scalings

We shall be working with the affine set Q defined in (1.2) and with the relative interior of its restriction to the first orthant S. We assume that S is nonempty. Given a point $\bar{x} \in \mathbb{R}^n_+$, we define a *scaling* about $\bar{x}$ as a change of coordinates $x = \bar{X}y$, where $\bar{X} = \operatorname{diag}(\bar{x}_1, \bar{x}_2, \ldots, \bar{x}_n)$. Notice that since $\bar{x} > 0$, the transformation is well defined, and it defines new norms $\|\cdot\|_{\bar{x}}$ for $\mathbb{R}^n$ such that for all $z \in \mathbb{R}^n$,

$$\|z\|_{\bar{x}} = \|\bar{X}^{-1}z\|,$$
$$\|z\|_{\bar{x}}^{\infty} = \|\bar{X}^{-1}z\|_{\infty}. \tag{1.4}$$

Given a point $x \in \mathbb{R}_+^n$ and the linear programming problem (1.1), scaling about x produces an equivalent problem

$$\begin{aligned} &\text{minimize } (Xc)'y \\ &\text{subject to } (AX)y = b, \\ &\qquad\qquad y \geq 0. \end{aligned} \tag{1.5}$$

For this problem $y = e$ corresponds to x. If $x \in S$, then $y = e$ is feasible for the scaled problem.

Now define $\bar{A} = AX, \bar{c} = Xc$. The scaled linear programming problem will then be

$$\begin{aligned} &\text{minimize } \bar{c}'y \\ &\text{subject to } \bar{A}y = b, \\ &\qquad\qquad y \geq 0. \end{aligned} \tag{1.6}$$

This problem has the same format as (1.1). We shall denote by $\bar{S}, \bar{Q}, \bar{D}$ the sets corresponding respectively to S, Q, D in (1.1).

In dealing with the scaled problem we will need the projections of the vectors $\bar{c}$ and e onto $\bar{D}$, and this is the most time-consuming operation of all necessary to algorithms based on scaling.

Scaling affects the steepest descent direction in an optimization algorithm, and it can be used to improve the performance of first-order feasible direction methods. This is the case with Karmarkar's algorithm, which computes at each iteration a steepest descent direction for his "logarithmic Potential function." On the other hand, the Newton-Raphson method is scale invariant, and no change in its performance can be gained by such coordinate changes. In our approach, scaling will be used for the purpose of simplifying the mathematical treatment.

Our first lemma asserts that scalings about points "near" each other have similar effects on the scaled norms. In this lemma proximity will be measured in the norm of the sup and is obviously valid for any other standard norm for $\mathbb{R}^n$.

Lemma 1.1. *Let $a, b \in \mathbb{R}_+^n$ be such that $\|a - b\|_a^{\infty} \leq 0.1$. Then given any $h \in \mathbb{R}^n$,*

$$0.9 \leq \frac{\|h\|_a}{\|h\|_b} \leq 1.1.$$

PROOF. Let

$$h_a = \left[\frac{h_i}{a_i}\right] = \left[\frac{h_i}{b_i}\frac{b_i}{a_i}\right].$$

Note that $\|h_a\| = \|h\|_a$, and similarly $\|h_b\| = \|h\|_b$. Consequently,

$$\|h_a\|^2 = \sum_{i=1}^{n} \left(\frac{h_i}{b_i}\right)^2 \left(\frac{b_i}{a_i}\right)^2 \geq \left(\frac{b_i}{a_i}\right)^2_{\min} \left\| \left[\frac{h_i}{b_i}\right] \right\|^2,$$

where $(b_i/a_i)_{\min} = \min_{i=1,\ldots,n} |b_i/a_i| \geq 1 - 0.1$, since $|1 - b_i/a_i| \leq 0.1$. It follows that

$$\|h_a\| \geq 0.9 \|h\|_b.$$

The other inequality is proved similarly, using

$$\|h_a\| \leq \left(\frac{b_i}{a_i}\right)_{\max} \|h\|_b.$$

In the lemma above we used for the first time the criterion $\|z - x\|_x < \delta$ to assert that the points x and z are "near" each other. This means that scaling the problem about x, the point corresponding to z will be near e.

Remark. The lemma can be adapted to other values of δ, and the following relation will be useful: in the conditions of Lemma 1.1, if $\|a - b\|_a^\infty \leq 0.02$, then

$$\|h\|_b \leq 1.02 \|h\|_a. \tag{1.7}$$

The next lemma shows that if measured by the cost function, "near" is really near.

Lemma 1.2. *Let $x, z \in S$ be such that $\|x - z\|_x \leq \delta < 1$ in problem (1.1), and let v^* be the cost of an optimal solution for the problem. Then $c'z - v^* \leq (1 + \delta)(c'x - v^*)$.*

Proof. Consider the problem scaled about x, as in (1.6). Define $z_x = X^{-1}z$. Then $\bar{c}'e = c'x$ and $\bar{c}'z_x = c'z$. Let

$$\alpha = \max_{h \in D} \{\bar{c}'h \mid \|h\| \leq 1\}.$$

Since $\{e + h \mid \|h\| < 1, h \in D\} \subset \bar{S}$,

$$v^* \leq \bar{c}'e - \alpha = c'x - \alpha, \quad \text{or}$$

$$\alpha \leq c'x - v^*.$$

On the other hand,

$$c'z = \bar{c}'z_x \leq \bar{c}'e + \delta\alpha = c'x + \delta\alpha.$$

Merging the last two inequalities,

$$c'z \leq c'x(1 + \delta) - \delta v^*.$$

Subtracting v^* from both sides,

$$c'z - v^* \leq (1 + \delta)(c'x - v^*),$$

completing the proof.

2.2. Barrier Functions

We shall study penalized functions for problem (1.1). We do not assume that the simplex constraint is present but assume instead that the set S is bounded. This includes problem (1.6) and thus covers all cases to appear. The logarithmic barrier function is defined as

$$x \in \mathbb{R}^n_+ \mapsto p(x) = -\sum_{i=1}^{n} \log x_i. \tag{1.8}$$

Given a positive constant ε, a *penalized* objective function is defined by

$$x \in \mathbb{R}^n_+ \mapsto f_\varepsilon(x) = c'x + \varepsilon p(x) = c'x - \varepsilon \sum_{i=1}^{n} \log x_i. \tag{1.9}$$

For any $\varepsilon > 0$, $f_\varepsilon(\cdot)$ is well defined, is analytic and strictly convex, and has a global minimum over S.

Penalized Problem

Given $\varepsilon > 0$, the penalized problem associated to the linear programming problem (1.1) is

$$\min_{x \in S} f_\varepsilon(x). \tag{1.10}$$

The Newton-Raphson Step

Consider the penalized problem (1.10) and a given point $x^0 \in S$. An N-R step for this problem is a direction $h_N \in D$ that solves the problem

$$\min_{h \in D} f_N(x^0 + h), \tag{1.11}$$

where $f_N(\cdot)$ is the quadratic approximation to $f_\varepsilon(\cdot)$, given by

$$h \in D \mapsto f_N(x^0 + h) = f_\varepsilon(x^0) + \nabla f_\varepsilon(x^0)'h + \tfrac{1}{2}h'\nabla^2 f_\varepsilon(x^0)h. \tag{1.12}$$

Both (1.10) and (1.11) are well defined, since $f_\varepsilon(\cdot)$ grows indefinitely near the boundary of S and $f_N(\cdot)$ is quadratic with positive definite Hessian, as we shall see below.

We have now listed all problems to be studied here. The LP problem (1.1) will be solved by a sequence of penalized problems (1.10). Approximate solutions for these will be found by solving (1.11). The last link in this chain of simplifications will be introduced later by allowing well-bounded errors in the computation of the Hessian matrices.

Lemma 1.3. *The penalized function* (1.9) *has the values and derivatives below, and* $\nabla^2 f_\varepsilon(x)$ *is positive definite for any* $x \in \mathbb{R}^n_+$.

$$
\begin{aligned}
f_\varepsilon(e) &= c'e, \\
\nabla f_\varepsilon(e) &= c - \varepsilon e, \qquad \nabla f_\varepsilon(x) = c - \varepsilon[x_i^{-1}], \\
\nabla^2 f_\varepsilon(e) &= \varepsilon I, \qquad \nabla^2 f_\varepsilon(x) = \varepsilon X^{-2}.
\end{aligned}
\tag{1.13}
$$

PROOF. Straightforward.

We must now examine how well the quadratic approximation to the penalized function approximates its actual values. The result comes as a consequence of properties of the logarithm function.

Lemma 1.4. *Given a direction* $h \in \mathbb{R}^n$ *such that* $\|h\| < 1$, *the barrier function* (1.8) *can be written as*

$$
\begin{aligned}
p(e + h) &= -e'h + \frac{\|h\|^2}{2} + o(h), \quad \text{where} \\
|o(h)| &\le \frac{\|h\|^3}{3(1 - \|h\|)}.
\end{aligned}
\tag{1.14}
$$

PROOF. We start by writing a quadratic approximation for the logarithm on the real line:

$$
\log(1 + \lambda) = \lambda - \frac{\lambda^2}{2} + \frac{\lambda^3}{3} - \cdots. \tag{1.15}
$$

The error of the quadratic approximation for $\lambda < 1$ is given by

$$
|\delta(\lambda)| = \left| \sum_{i=3}^{\infty} \frac{\lambda^i(-1)^{i+1}}{i} \right| \le \sum_{i=3}^{\infty} \frac{|\lambda|^i}{3} = \frac{|\lambda|^3}{3(1 - |\lambda|)}.
$$

If we now consider a direction h with $\|h\| < 1$, and consequently $|h_i| < 1$ for $i = 1, 2, \ldots, n$,

$$
\begin{aligned}
p(e + h) &= -\sum_{i=1}^{n} \log(1 + h_i) \\
&= -\sum_{i=1}^{n} h_i + \sum_{i=1}^{n} \frac{h_i^2}{2} + \sum_{i=1}^{n} \delta(h_i) \\
&= -h'e + \frac{\|h\|^2}{2} + o(h),
\end{aligned}
$$

where

$$
|o(h)| \le \sum_{i=1}^{n} \frac{|h_i|^3}{3} \frac{1}{1 - |h_i|} \le \frac{1}{1 - \|h\|} \sum_{i=1}^{n} \frac{|h_i|^3}{3}.
$$

But $\sum_{i=1}^{n} |h_i|^3 = \|h\|_3^3 \le \|h\|^3$, by the ordering of norms in $\mathbb{R}^n$, and this completes the proof.

This lemma gives us good bounds on the third-order errors for the barrier function about the point e. The corresponding errors for the penalized functions (1.9) will then be given by $-\varepsilon o(h)$. Our next step is to estimate how well a minimizer of the quadratic approximation approximates the minimizer of the penalized function.

To simplify the notation, we shall work with the function

$$x \in \mathbb{R}^n, x > 0 \mapsto g(x) = \gamma' x + p(x) = f_\varepsilon(x)/\varepsilon. \tag{1.16}$$

The function $g(\cdot)$ is equivalent to $f_\varepsilon(\cdot)$ up to a multiplicative constant. Using (1.13), define its quadratic approximation about e as

$$h \in \mathbb{R}^n, e + h > 0 \mapsto g_N(e + h) = \gamma'(e + h) - e'h + \tfrac{1}{2}\|h\|^2, \tag{1.17}$$

By Lemma 1.4, for $e + h \in S$,

$$g(e + h) = g_N(e + h) + o(h).$$

Let the solution to the penalized problem and the N-R step be defined as usual by

$$\begin{aligned} \hat{x} &= e + \hat{h} = \operatorname{argmin}\{g(x) | x \in S\}, \\ x_N &= e + h_N = \operatorname{argmin}\{g_N(x) | x \in Q\}. \end{aligned} \tag{1.18}$$

The following lemma asserts that when $\mathbf{x}$ approaches e, the precision with which x_N estimates $\mathbf{x}$ increases.

Lemma 1.5. *Consider $\hat{h}$ and h_N defined as above, and let $\alpha > 0$ be given. Then there exists $\gamma > 0$ such that if either $\|\hat{h}\| \le \gamma$ or $\|h_N\| \le \gamma$ then $\|\hat{h} - h_N\| \le \alpha\gamma$.*

PROOF. We shall study the limitation in $\|\hat{h} - h_N\|$ for small values of $\hat{h}$ and h_N.

First part: assume that $\hat{h} \le 0.1$. Since we cannot assume that h_N is also small (although this is true), let us examine points in the segment joining $\hat{x}$ and x_N. Since $g_N(\cdot)$ is minimized at x_N and $\nabla^2 g_N(x_N) = I$, $g_N(\cdot)$ restricted to this segment is a parabola with minimum at x_N, and for any point z on this segment,

$$g_N(\hat{x}) \ge g_N(z) + \tfrac{1}{2}\|\hat{x} - z\|^2. \tag{1.19}$$

Choose such a point z satisfying $\|\hat{x} - z\| \le \|\hat{h}\|$ and define $d = z - \hat{x}$. Then, using Lemma 1.4,

$$\begin{aligned} g_N(\hat{x}) &\le g(\hat{x}) + |o(\hat{h})| \\ &\le g(z) + |o(\hat{h})| \quad \text{by definition of } \hat{x}, \\ &\le g_N(z) + |o(\hat{h})| + |o(\hat{h} + d)|. \end{aligned} \tag{1.20}$$

Subtracting (1.19) from (1.20),

$$\tfrac{1}{2}\|d\|^2 \le |o(\hat{h})| + |o(\hat{h} + d)|$$

$$\le \frac{\|\hat{h}\|^3}{3(1 - \|\hat{h}\|)} + \frac{\|\hat{h} + d\|^3}{3(1 - \|\hat{h} + d\|)}. \tag{1.21}$$

Now define μ by $\|d\| = \mu\|\hat{h}\|$, $\mu \le 1$. Since the denominators in (1.21) are greater than 0.8,

$$\frac{1}{2}\mu^2 \le \frac{1}{2.4}[1 + (1 + \mu)^3]\|\hat{h}\| \le \frac{9}{2.4}\|\hat{h}\|,$$

since $\mu < 1$. We conclude that μ decreases with $\sqrt{\|\hat{h}\|}$, completing the first part of the proof.

Second part: we must now consider the case in which $\|h_N\|$ is known to be small. The proof is identical, except for the choice of z near x_N instead of near $\hat{x}$. All relations above are then satisfied for z substituting $\hat{x}$ and x_N substituting z.

It is interesting to check some numerical values for expression (1.21). For $\|\hat{h}\|$ very small, the relationship tends to $\mu^2 \le (4/3)\|\hat{h}\|$, or $\|d\| \le 1.16\|\hat{h}\|^{1.5}$, obtained by taking limits. The following values will prove useful and can be directly verified:

Lemma 1.6. *Let $\hat{x}$ and x_N be defined as in* (1.18). *Then*:

$$\textit{If } \|x_N - e\| \le 0.02 \textit{ then } \|\hat{x} - x_N\| \le 1.4\|x_N - e\|^{1.5} \le 0.004.$$

$$\textit{If } \|\hat{x} - e\| \le 0.025 \textit{ then } \|\hat{x} - x_N\| \le 0.005.$$

2.3. Computation of a Newton-Raphson Step

We now show how to solve for the N-R step in (1.11). Let P be the projection matrix onto D, and set $c_p = Pc$, $e_p = Pe$.

Lemma 1.7. *Consider the N-R step problem* (1.11) *for given $\varepsilon > 0$, $x^0 \in S$. The N-R step h_N satisfies*

$$PX_0^{-2}h_N = -\frac{c_p}{\varepsilon} + PX_0^{-1}e. \tag{1.22}$$

In particular, for $x^0 = e$,

$$h_N = -\frac{1}{\varepsilon}P\nabla f_\varepsilon(e) = -\frac{c_p}{\varepsilon} + e_p. \tag{1.23}$$

Proof. As we saw before, h_N is well defined. At $x^0 + h_N$ we must have

$$P\nabla f_N(x^0 + h_N) = 0$$

or, since f_N is quadratic,

$$P(\nabla f_\varepsilon(x^0) + \nabla^2 f_\varepsilon(x^0) h_N) = 0. \tag{1.24}$$

Using (1.13),

$$P(c - \varepsilon X_0^{-1} e + \varepsilon X_0^{-2} h_N) = 0.$$

This expression is equivalent to the desired result. For $x_0 = e$, $X_0 = I$, (1.22) becomes

$$Ph_N = -\frac{c_p}{\varepsilon} + e_p.$$

The proof is completed by noting that since $h_N \in D$, $Ph_N = h_N$.

Expression (1.23) resumes what is well known in nonlinear programming: scaling about x^0 corresponds to a change of metric in which the steepest descent coincides with the N-R direction. In other words, the change of coordinates corresponding to the scaling operation transforms the ellipsoidal level sets for $f_N(\cdot)$ into spheres.

Approximate Hessians

Our last result in this section will be an account of the effect of errors in the Hessian. We shall consider errors of the following kind: given $x^0 \in S$ and $\tilde{x}$ near x^0 ($\tilde{x}$ not necessarily in S), the N-R iteration from x^0 will be computed using the wrong values $\nabla^2 f_\varepsilon(x^0) \leftarrow \varepsilon \tilde{X}^{-2}$ instead of $\nabla^2 f_\varepsilon(x^0) = \varepsilon X_0^{-2}$. In terms of the algorithm to be studied here, this amounts to not updating the Hessian matrices in every iteration: only entries with a large error will be changed. This will give rise to rank-1 updates in the solution of the N-R steps instead of full matrix inversions.

Note that saving time by approximating the Hessians is current practice in predictor-corrector algorithms, and again we are using traditional procedures.

It is easy to see that by scaling the problem, it is enough to consider the case $\tilde{x} = e$, and the error in the Hessian reduces to $\nabla^2 f_\varepsilon(x^0) \leftarrow \varepsilon I$. This has a slightly different interpretation: instead of rescaling the problem at every iteration about x^k, the scaling is done about a point near x^k. This will be discussed in depth in Section 4.

The next lemma relates errors in the Hessian to errors in the resulting N-R step.

Lemma 1.8. *Consider the N-R step problem* (1.11) *for given* $\varepsilon > 0$, $x \in S$. *Let* h_N *be the N-R step computed by* (1.22) *and define*

$$\tilde{h}_N = -\frac{1}{\varepsilon} P\nabla f_\varepsilon(x) = -\frac{c_p}{\varepsilon} + PX^{-1}e. \tag{1.25}$$

If $\|x - e\|_\infty \leq 0.1$ *then* $\|h_N - \tilde{h}_N\| \leq 0.235 \|h_N\|$.

PROOF. Using (1.24),

$$P\nabla f_\varepsilon(x) + \varepsilon P X^{-2} h_N = 0, \quad \text{or}$$

$$P X^{-2} h_N = -\frac{1}{\varepsilon}\nabla f_\varepsilon(x) = \tilde{h}_N.$$

But by hypothesis for $i = 1, \ldots, n$

$$x_i^{-2} = \frac{1}{(1 + v_i)^2} \quad \text{where } |v_i| \le 0.1,$$

and consequently, as can easily be verified,

$$x_i^{-2} = 1 + \mu_i \quad \text{where } |\mu_i| \le 0.235.$$

Setting $M = \operatorname{diag}(\mu_i)$, it follows that

$$\tilde{h}_N = P X^{-2} h_N = P I h_N + P M h_N = h_N + P M h_N,$$

Consequently,

$$\|h_N - \tilde{h}_N\| = \|P M h_N\| \le \|M h_N\| \le 0.235\|h_N\|,$$

completing the proof.

To finish this section, we shall state a new version of the result above, in a format that will be shown to be useful.

Lemma 1.9. *Consider the N-R step problem* (1.11) *for given* $\varepsilon > 0$, $z^k \in S$, *and let* h_N *be the N-R step computed by* (1.22). *Let* $\tilde{z} \in \mathbb{R}^n_+$ *be a point such that* $\|z^k - \tilde{z}\|_{\tilde{z}}^{\infty} \le 0.1$, *and define* $\tilde{h}_N$ *as the result of the N-R step problem with Hessian* Z_k^{-1} *approximated by* $\tilde{Z}^{-1}$. *Then*

$$\|h_N - \tilde{h}_N\|_{z^k} \le 0.29\|h_N\|_{z^k}.$$

PROOF. Scaling about $\tilde{z}$, the situation in Lemma 1.8 is reproduced. The resulting directions satisfy

$$\|h_N - \tilde{h}_N\|_{\tilde{z}} \le 0.235\|h_N\|_{\tilde{z}}.$$

To translate the result to $\|\cdot\|_{z^k}$, all we need is to use Lemma 1.1 on both sides of the equation, obtaining

$$\|h_N - \tilde{h}_N\|_{z^k} \le 0.235 \times 1.1 \times 1.1\|h_N\|_{z^k} \le 0.29\|h_N\|_{z^k}.$$

completing the proof.

We are now well equipped to study the barrier function method and its approximations.

§3. The Conceptual Algorithm

Consider the linear programming problem (1.1). The conceptual barrier function method will use penalty multipliers $\varepsilon_k = (1 - \sigma)^k \varepsilon_0$, where $\varepsilon_0 > 0$ and $\sigma \in (0, 1)$ are given. Each iteration solves exactly a minimization problem with criterion $f_k(x) = c'x + \varepsilon_k p(x)$. The algorithm iterates until ε_k falls under a given precision $\delta > 0$. Notice that this is equivalent to fixing the number of iterations to a number K such that $\varepsilon_0(1 - \sigma)^K < \delta$.

Algorithm 1.1. Conceptual barrier function: given $\varepsilon_0 > 0$, $\delta > 0$, $\sigma \in (0, 1)$.
$k := 0$
Repeat
  Compute a solution x^k to the problem

$$\min_{x \in S} \left(c'x - \varepsilon_k \sum_{i=1}^{n} \log x_i \right) \qquad (P_k)$$

  $\varepsilon_{k+1} := (1 - \sigma)\varepsilon_k$
  $k := k + 1$
Until $\varepsilon_k < \delta$.

This section is dedicated to defining "small steps" for the conceptual algorithm 1.1 and finding a value of the penalty adaptation parameter σ that guarantees such small steps.

3.1. Convergence of the Conceptual Algorithm

Let us begin by describing properties of the conceptual algorithm. Consider problem (1.1), and let x^* be an optimal solution and call its value $v^* = c'x^*$.

Lemma 1.10. *The sequence* $(x^k)_{k \in N}$ *generated by Algorithm* 1.1 *satisfies:*

(i) *For any given* $x \in S$, $f_k(x) \to c'x$.
(ii) $f_k(x^k) \to v^*$.

PROOF. (i) is immediate, since for any $x \in S$, $f_k(x) = c'x + \varepsilon_k p(x)$, and $\varepsilon_k \to 0$. To prove (ii), assume by contradiction that for all $k \in M \subset N$, $f_k(x^k) \geq v^* + 2\delta$, $\delta > 0$, where M is an infinite set.

Since any optimal solution is in the closure of S, there exists $x \in S$ such that $c'x < v^* + \delta$. It follows that for all $k \in M$,

$$f_k(x) \geq f_k(x^k) \geq v^* + 2\delta \geq c'x + \delta.$$

This contradicts (i) and completes the proof.

Lemma 1.11. *Let* x^k *be a point generated by Algorithm* 1.1. *Then*

$$c'x^k - v^* \leq n\varepsilon_k.$$

PROOF. From (1.13), $\nabla f_k(x^k) = c - \varepsilon_k X_k^{-1}e$. Since x^k solves problem (P_k), $\nabla f_k(x^k)$ is orthogonal to any feasible direction. In particular, for the direction $h = x^k - x^*$,

$$\nabla f_k(x^k)'(x^k - x^*) = c'x^k - c'x^* + \varepsilon_k e'X_k^{-1}x^* - \varepsilon_k e'X_k^{-1}x^k = 0.$$

But $X_k^{-1}x^k = e$, and then

$$c'x^k - v^* = \varepsilon_k(n - e'X_k^{-1}x^*).$$

Since $x^* \geq 0$ and $x^k > 0$, it follows that $e'X_k^{-1}x^* \geq 0$, completing the proof.

Lemma 1.11 links the improvement in cost to the evolution of the sequence (ε_k). This means that costs $c'x^k$ converge to v^* at the speed with which ε_k approaches zero. Our task is then to reduce these multipliers as fast as possible while keeping our short-step strategy.

3.2. Differential Properties: The Homotopy Approach

In the algorithm above we made an option for the barrier function approach that will be kept in the remainder of the chapter. We pause now to make informal comments on the alternative homotopy framework.

If we define the homotopy mapping

$$x \in S,\ \varepsilon > 0 \mapsto H(x, \varepsilon) = P\nabla_x f_\varepsilon(x),$$

the path of solutions to the penalized problem is defined by the homotopy

$$H(x, \varepsilon) = 0, \qquad x \in S,\ \varepsilon > 0.$$

It is easy to derive differential properties for the solution path, following reference [4]. Considering the path parametrized by ε, the homotopy equation $H(x(\varepsilon), \varepsilon) = 0$ can be differentiated, resulting in

$$\frac{\partial H}{\partial x}\frac{dx}{d\varepsilon} + \frac{\partial H}{\partial \varepsilon} = 0, \qquad \frac{\partial H}{\partial x} = \varepsilon P X^{-2}, \qquad \frac{\partial H}{\partial \varepsilon} = -PX^{-1}e.$$

Calculating these terms for $x = e$, we obtain

$$\varepsilon P \frac{dx}{d\varepsilon} = e_p.$$

But for $x(\varepsilon) \in S$,

$$P\frac{dx}{d\varepsilon} = \frac{dx}{d\varepsilon}, \quad \text{and}$$

$$\frac{dx}{d\varepsilon} = \frac{1}{\varepsilon} e_p.$$

This can be written as

$$\Delta x = \frac{\Delta\varepsilon}{\varepsilon} e_p + o(\Delta\varepsilon),$$

where $o(\Delta\varepsilon)$ is an error. Taking norms, it follows that

$$\|\Delta x\| \leq \frac{|\Delta\varepsilon|}{\varepsilon}\sqrt{n} + \|o(\Delta\varepsilon)\|. \tag{1.26}$$

This is a very important result. It means that if we choose a small α and at each iteration set

$$\frac{\Delta\varepsilon}{\varepsilon} \leq \frac{\alpha}{\sqrt{n}},$$

then $\|\Delta x\| \leq \alpha + \|o(\Delta\varepsilon)\|$.

In other words, a small variation in x is associated with a variation in the penalty parameter (and thus in the objective) that depends on $\sqrt{n}$. Small variations in x mean good precision for the N-R steps; improvements dependent on $\sqrt{n}$ will lead us to the solution of the LP problem in $O(n^{0.5}L)$ iterations.

Unfortunately, complexity studies need precise bounds on $o(\Delta\varepsilon)$, not provided by the analysis above. The rest of this section will be used to derive this result again, without taking any limits. The barrier function approach seems more adapted to this purpose.

3.3. The Choice of ε_k

We saw in last section that the precision with which a Newton-Raphson iteration can approximate the minimizer of a penalized function starting from the point e depends only on how far the minimizer is from e. This gives us the clue to the small step: two consecutive minimizers x^k, x^{k+1} will be considered as near each other if $\|x^k - x^{k+1}\|_{x^k}$ is small. In other words, we want the solution of each penalized problem to be near e when solving the problem scaled about x^k.

Assume that in iteration k of the conceptual algorithm 1.1 we know x^k exactly; then set $\varepsilon_{k+1} = (1-\sigma)\varepsilon_k$ and use an N-R iteration to find an approximation to x^{k+1}.

Instead of performing this in the original space, we shall scale the problem about x^k, obtaining problem (1.6). In this problem the vector e is feasible and corresponds to x^k in the original coordinates.

Objective Function

The penalized function for the scaled problem will be

$$y \in \bar{S} \mapsto g(y) = \bar{c}'y - \varepsilon_{k+1} \sum_{i=1}^{n} \log y_i. \tag{1.27}$$

Lemma 1.12. *The penalized function is scale invariant in the sense that for any $y \in \bar{S}$, $g(y) = f_{k+1}(X_k y) + K$, where K is constant with y.*

PROOF. Developing $f_{k+1}(\cdot)$,

$$\begin{aligned} f_{k+1}(X_k y) &= c' X_k y - \varepsilon_{k+1} \sum_{i=1}^{n} \log x_i^k y_i \\ &= (X_k c)' y - \varepsilon_{k+1} \sum_{i=1}^{n} \log y_i - \varepsilon_{k+1} \sum_{i=1}^{n} \log x_i^k. \end{aligned}$$

But $(X_k c)' y = \bar{c}' y$, and then setting $K = -\varepsilon_{k+1} \sum_{i=1}^{n} \log x_i^k$ completes the proof.

This shows that minimizing (1.27) in $\bar{S}$ is equivalent to minimizing $f_{k+1}(\cdot)$ in S. We now proceed with the scaled problem.

Let $r(\cdot)$ be defined as the penalized function in $\bar{S}$ with penalty multiplier ε_k:

$$y \in \bar{S} \mapsto r(y) = \bar{c}' y - \varepsilon_k \sum_{i=1}^{n} \log y_i. \tag{1.28}$$

Applying Lemma 1.12, this function assumes a minimum over $\bar{S}$ at e, the point corresponding to x^k.

Setting $\varepsilon_{k+1} = (1 - \sigma)\varepsilon_k$ and computing derivatives by (1.13),

$$\begin{aligned} g(e) &= r(e) = \bar{c}' e, \\ \nabla r(e) &= \bar{c} - \varepsilon_k e, \\ \nabla g(e) &= \bar{c} - (1 - \sigma)\varepsilon_k e = \nabla r(e) + \sigma \varepsilon_k e. \end{aligned} \tag{1.29}$$

We are ready for the most important result, which, like (1.26), relates σ to the distance from e to the minimizer of $g(\cdot)$. This will be obtained in two steps: we first deal with the minimizer of $g_N(\cdot)$ and then extend the result to the minimizer of $g(\cdot)$.

Lemma 1.13. *Let $\delta \in (0, 0.1)$ be a given constant. If $\sigma \leq \delta/\sqrt{n}$, then the N-R step h_N for $g(\cdot)$ from e satisfies $\|h_N\| \leq \delta$.*

PROOF. Setting $\varepsilon = \varepsilon_{k+1}$ in the expression for the N-R step (1.23),

$$h_N = -\frac{1}{\varepsilon_{k+1}} P \nabla g(e).$$

Taking the scalar product with h_N and noting that since h_N is feasible, $h_N' P y = h_N' y$ for any vector y,

$$\|h_N\|^2 = -\frac{1}{\varepsilon_{k+1}} h_N' \nabla g(e). \tag{1.30}$$

Now, using (1.29),

$$h'_N \nabla g(e) = \sigma \varepsilon_k e' h_N + h'_N \nabla r(e).$$

But e minimizes $r(\cdot)$ over $\bar{S}$, and then for the feasible direction $h_N \in D$, $h'_N \nabla r(e) = 0$. It follows that, merging the last equation into (1.30),

$$\|h_N\|^2 = -\sigma \frac{\varepsilon_k}{\varepsilon_{k+1}} e' h_N = -\frac{\sigma}{1-\sigma} e' h_N.$$

Now, using the well-known relationship between norms 1 and 2,

$$-e' h_N \leq \|h_N\|_1 \leq \sqrt{n} \|h_N\|.$$

It follows that

$$\|h_N\| \leq \frac{\sigma}{1-\sigma} \sqrt{n}.$$

If we choose $\sigma \leq \delta/\sqrt{n} \leq 0.08$ for $n > 1$,

$$\|h_N\| \leq 1.1 \sigma \sqrt{n} \leq \delta,$$

completing the proof.

Lemma 1.13 is almost what we need. To relate σ to the minimizer $\hat{y}$ of $g(\cdot)$ over $\bar{S}$ all we have to do is merge Lemmas 1.13 and 1.6. We shall do that for numerical values that will prove convenient later; it does not make sense to try to push the step size to its maximum, since it does not affect the speed of convergence and the maximum provable step size is small anyway.

Theorem 1.1. *Let the penalty multipliers in* 1.1 *be adapted by* $\varepsilon_{k+1} = (1-\sigma)\varepsilon_k$, *with* $\sigma = \delta/\sqrt{n}$. *If* $\delta \leq 0.005$, *then for all* $k \in N$, $\|x^k - x^{k+1}\|_{x^k} \leq 0.006$.

PROOF. Consider the problem scaled about x^k as above. Assuming $\sigma \leq 0.005/\sqrt{n}$, then by Lemma 1.13, $\|h_N\| \leq 0.005$.

Now Lemma 1.6 deals with the same scaled problem (with an objective function equivalent up to a product by a constant), and then

$$\|\hat{y} - e\| \leq \|h_N\| + \|\hat{y} - y_N\| \leq \|h_N\| + 1.4\|h_N\|^{1.5} \leq 0.006,$$

and the proof is complete.

Complexity of the Conceptual Algorithm

At this point we can advance that it will be possible to choose an initial penalty multiplier ε_0 bounded by 2^L. It is then immediate to compute the number of iterations necessary to obtain a reduction $\varepsilon_k \leq \varepsilon_0 2^{-L}$.

Lemma 1.14. *If the conceptual barrier function algorithm* 1.1 *uses an adaptation parameter* $\sigma = \delta/\sqrt{n}$, *then for* $k \geq \sqrt{n}L/\delta$, $\varepsilon_k \leq \varepsilon_0 2^{-L}$.

PROOF. Let $k \geq \sqrt{n}L/\delta$. Taking logarithms,

$$\begin{aligned}\log \varepsilon_k &= \log \varepsilon_0 + k \log(1 - \delta/\sqrt{n}) \\ &\leq \log \varepsilon_0 - k\delta/\sqrt{n} \quad \text{by (15)} \\ &\leq \log \varepsilon_0 - L.\end{aligned}$$

It follows that $\varepsilon_k \leq \varepsilon_0 \exp(-L) \leq \varepsilon_0 2^{-L}$, completing the proof.

In the next section we shall prove how to obtain the same speed of convergence with approximate computations; the total number of operations will then be given by the effort per iteration multiplied by $O(\sqrt{n}L)$.

§4. The Implementable Algorithm

The implementable algorithm uses approximate Newton-Raphson searches instead of exact minimizations for the penalized functions. The points x^k will never be computed; a sequence (z^k) will be computed instead, and we will make sure that each z^k is near the corresponding x^k. A convenient definition of "near" will be $\|z^k - x^k\|_{x^k} \leq 0.015$.

We begin by showing how to compute an initial penalty parameter ε_0 such that e (the initial point for the implementable algorithm) is near x^0 (the initial point for the conceptual algorithm). Then we develop a simplified algorithm model without specifying how to approximate the Hessian matrix and prove its efficiency in solving the linear programming problem. The final step will be to specify the details on Hessian approximations. We shall use the notation described in Sections 1 and 2.

As we pointed out before, scaling operations simplify the mathematical treatment but are not necessary at all. The algorithm model will use scalings for this reason. It is easy to modify the algorithm so that it does not rely on scalings, and we shall indicate how to do it immediately after presenting the algorithm.

The Initial Penalty Multiplier

Consider problem (1.1) and the penalized problems introduced in (1.3) and studied in Section 2. Assume that the simplex constraint is enforced and that e is feasible. A procedure for achieving this was described by Karmarkar [8]. Note that this is the only point in the chapter in which this constraint is used. It can also be of interest to note that we need only know the point of minimum penalty in S, trivially equal to e if the simplex constraint is enforced.

Lemma 1.15. *Consider problem* (P_0), *in the first iteration of the conceptual algorithm* 1.1. *If* $\varepsilon_0 \geq \|c\|/0.01$ *then* $\|e - x^0\|_{x^0} \leq 0.015$.

PROOF. In the first iteration, the N-R direction from e is computed by (1.23):

$$h_N = -\frac{c_p}{\varepsilon_0} + e_p.$$

But in the original linear programming problem the simplex constraint is in use and then $e_p = Pe = 0$. Eliminating this term and taking the scalar product with h_N,

$$\|h_N\|^2 = -\frac{h'_N c_p}{\varepsilon_0} \leq \frac{\|c_p\|}{\varepsilon_0} \|h_N\|.$$

Since $\|c_p\| \leq \|c\|$, it follows that

$$\|h_N\| \leq \frac{\|c\|}{\varepsilon_0}.$$

Now using $\varepsilon_0 \geq \|c\|/0.01$,

$$\|h_N\| \leq 0.01.$$

Using Lemma 1.6, with $\hat{x} = x^0$ and $x_N = e + h_N$,

$$\|x^0 - e\| \leq \|h_N\| + \|x^0 - x_N\| \leq 0.01 + 0.002 \leq 0.012.$$

Since $\|e - x^0\| \leq 0.09$, we can use Lemma 1.1, obtaining

$$\|x^0 - e\|_{x^0} \leq 1.1 \|x^0 - e\|_e \leq 0.015$$

since $\|x^0 - e\|_e = \|x^0 - e\|$, and this completes the proof.

The Algorithm Model

Consider the linear programming problem (1.1). Each iteration will perform a scaling operation, define a penalized function, and perform a Newton-Raphson search. Instead of computing the exact N-R step, a point $\tilde{z}$ near the present iterate z^k will be *chosen* (this is the reason for calling this a model; the determination of $\tilde{z}$ will be done later), and the Hessian at z^k will be approximated by the Hessian at $\tilde{z}$. As we saw in Lemma 1.9, this is equivalent to a steepest descent step for the problem scaled about $\tilde{z}$ instead of z^k.

Algorithm 1.2. Model: given $\sigma = 0.005/\sqrt{n}$, $\delta > 0$.

$k := 0$, $z^0 := e$

$\varepsilon_0 := \|c\|/0.01$

Repeat

$\varepsilon_{k+1} = (1 - \sigma)\varepsilon_k$

Choose $\tilde{z}$ such that $\|\tilde{z} - z^k\|_{\tilde{z}}^{\infty} \leq 0.1$

Scale the problem about $\tilde{z}$, obtaining problem (1.6), and define the penalized problem

$$\min_{y \in S} \left(\bar{c}'y - \varepsilon_{k+1} \sum_{i=1}^{n} \log y_i \right).$$

Compute the projection matrix $\bar{P}$ onto the feasible set and set

$$y := \tilde{Z}^{-1}z^k, \qquad Y := \operatorname{diag}(y_i).$$

Compute an approximated N-R step from y, obtaining by (1.25),

$$\tilde{h}_N = -\frac{\bar{P}\bar{c}}{\varepsilon_{k+1}} + \bar{P}Y^{-1}e.$$

$\tilde{y}_N := y + \tilde{h}_N$
Return to the original space with $z^{k+1} = \tilde{Z}\tilde{y}_N$
$k := k + 1$
Until $\varepsilon_k < \delta$

The algorithm can be modified so as not to mention scaling operations simply by saying "compute an approximated N-R step from z^k with Hessian approximated by $\tilde{Z}^{-2}$." In this case the rank-1 updates in the N-R step equations must be worked out. We shall not do it.

Now consider the sequence (x^k) that would be generated by the conceptual algorithm with the same penalty multipliers. The theorem to follow is the main result in this section.

Theorem 1.2. *At any iteration k of Algorithm* 1.2, *the following relations are satisfied*:

$$\|z^k - x^k\|_{x^k} \leq 0.015,$$

$$\|z^k - z^{k+1}\|_{z^k} \leq 0.04.$$

PROOF. The proof is done by induction. By Lemma 1.15,

$$\|z^0 - x^0\|_{x^0} \leq 0.015.$$

Assume that at iteration k

$$\|z^k - x^k\|_{x^k} \leq 0.015.$$

By Theorem 1.1, since $\sigma = 0.005/\sqrt{n}$,

$$\|x^{k+1} - x^k\|_{x^k} \leq 0.006.$$

Adding the two last inequalities,

$$\|z^k - x^{k+1}\|_{x^k} \leq 0.021.$$

By (1.7),

$$\|z^k - x^{k+1}\|_{z^k} \leq 0.022. \tag{1.31}$$

Now consider the (unknown) exact solution of an N-R step $z_N = z_k + h_N$ and the approximate solution $z^{k+1} = z^k + \tilde{h}_N$ found by the algorithm. By Lemma 1.6,

$$\|z_N - x^{k+1}\|_{z^k} \leq 0.005. \tag{1.32}$$

Adding the two last inequalities,

$$\|h_N\|_{z^k} \leq \|z_N - x^{k+1}\|_{z^k} + \|x^{k+1} - z^k\|_{z^k} \leq 0.027.$$

We can now use Lemma 1.9 to obtain

$$\|z^{k+1} - z_N\|_{z^k} = \|h_N - \tilde{h}_N\|_{z^k} \leq 0.29 \times 0.027 \leq 0.008.$$

Adding this inequality to (1.32),

$$\|z^{k+1} - x^{k+1}\|_{z^k} \leq 0.013. \tag{1.33}$$

Finally, again using Lemma 1.1,

$$\|z^{k+1} - x^{k+1}\|_{x^{k+1}} \leq 0.015,$$

This proves the first relationship. The second one is immediately obtained by adding (1.31) and (1.33), completing the proof.

The implementable algorithm generates a sequence (z^k), each z^k near the corresponding conceptual x^k. By Lemma 1.2, the costs of these pairs of vectors differ by a negligible amount. We are finally ready to extend the property found for the conceptual algorithm in Section 3 to any algorithm in this model.

Theorem 1.3. *Consider Algorithm 1.2 applied to the linear programming problem defined in Section 1, with $\delta \leq 2^{-L}/1.015n$, where L is the total length of the input data. Then the algorithm terminates in $O(\sqrt{n}L)$ iterations with a feasible solution z^k such that $c'z^k - v^* \leq 2^{-L}$.*

PROOF. The initial penalty multiplier satisfies $\varepsilon_0 = \|c\|/0.01 \leq 100 \times 2^L$. The algorithm stops at iteration k such that $\varepsilon_k \leq \delta$, or

$$\frac{\varepsilon_k}{\varepsilon_0} \leq \frac{2^{-2L}}{1.015n}.$$

This reduction is of the order $O(2^L)$ and can be achieved in $O(\sqrt{n}L)$ iterations, by Lemma 1.14.

Applying Lemmas 1.2 and 1.11 to the final solution generated by the algorithm,

$$c'z^k - v^* \leq 1.015(c'x^k - v^*) \leq 1.015n\varepsilon_k \leq 1.015n\delta \leq 2^{-L}.$$

completing the proof.

The choice of $\tilde{z}$

An obvious choice for $\tilde{z}$ is $\tilde{z} = z^k$. This choice causes one projection computation per iteration, with a bound of $O(n^3)$ operations. The algorithm then terminates with a bound of $O(n^{3.5}L)$ operations, as in the methods by Karmarkar and Renegar.

We shall lower this bound by saving in the projection computations. The intuitive argument is as follows: each iteration causes a move (in the scaled problem) of $\|\delta\|$; each approximation for the Hessian is good in a "cubic" ball $\|w - y\|_\infty \le 0.1$. Since the volume of a ball in the sup norm is related to the volume of a ball in the Euclidean norm by a factor of $\sqrt{n}$, it is reasonable to expect that this will be the relationship between the number of iterations and the number of complete recalculations of the projection matrix. This feature will be explored in the procedure below, in which a component of $\tilde{z}$ is changed only when it is too far from z^k.

In the first iteration, set $\tilde{z} := e$ and compute the projection matrix P by any convenient method. For $k > 0$, use the algorithm below.

Algorithm 1.3. Updates of $\tilde{z}$ and $\bar{P}$

Set

$$J := \left\{ j = 1, \ldots, n \,\middle|\, \frac{|z_j^k - \tilde{z}_j|}{\tilde{z}_j} \ge 0.1 \right\}.$$

For all $j \in J$ set

$$\tilde{z}_j := z_j^k.$$

Compute the projection matrix $\bar{P}$ from the former one by $|J|$ rank-1 updates.

The projection matrix for each scaled problem is calculated by $\bar{P} = I - \tilde{Z}A'(A\tilde{Z}^2A')^{-1}A\tilde{Z}$. The main effort is in computing $(A\tilde{Z}^2A')^{-1}$. If only one entry of $\tilde{z}$ changes, this computation can be performed in $O(n^2)$ operations by a rank-1 update, as was established in [8]. This fact will be used in the next section to study the complexity of the algorithm.

§5. Complexity of the Algorithm

Consider Algorithm 1.2 with $\tilde{z}$ chosen as in 1.3. We shall now establish an upper bound on the total number of rank-1 updates computed by the algorithm. The updates are similar to the ones in Karmarkar's algorithm, but our proofs follow a different path.

Consider the sequence $(z^k)_{k=0,\ldots,K+1}$ generated by the algorithm, where K is the index of the last iteration, and the associated real sequences (z_j^k) for each component $j = 1, \ldots, n$. We begin by analyzing a single component j.

To simplify the notation, let us denote $w \equiv z_j$ and deal with the sequence $(w^k) \equiv (z_j^k)$. Define the subsequence $(w^{k_i})_{k_i \in T_j}$ corresponding to the iterations in which $\tilde{z}_j$ is updated. T_j is the set of indices of such iterations, and its cardinality $|T_j|$, the total number of updates, is the object of our study.

Now consider two consecutive indices $k_i, k_{i+1} \in T_j$. The following facts are true by construction:

$$\left| \frac{w^{k_{i+1}}}{w^{k_i}} - 1 \right| \ge 0.1,$$

since at iteration k_{i+1} procedure 1.3 updates $\tilde{z}_j$. Between two updates,

$$\text{for } k_i \leq k < k_{i+1}, \qquad \left|\frac{w^k}{w^{k_i}} - 1\right| \leq 0.1.$$

Taking logarithms and defining $\mu = \min\{\log 1.1 - \log 0.9\} > 0.09$, we obtain for any two consecutive indices in T_j,

$$|\log w^{k_i} - \log w^{k_{i+1}}| \geq \mu. \tag{1.34}$$

The next lemma is preparatory for the main result to follow.

Lemma 1.16. *For the sequence constructed above, for a component* $j = 1, \ldots, n$,

$$|T_j| \leq \frac{1.1}{\mu} \sum_{k=0}^{K} \frac{|w^{k+1} - w^k|}{w^k}. \tag{1.35}$$

PROOF. Consider initially any two indices $0 \leq k_1 < k_2 \leq K$. Then, manipulating the integral of the logarithmic function,

$$|\log w^{k_1} - \log w^{k_2}| = \left|\int_{w^{k_1}}^{w^{k_2}} \frac{dw}{w}\right|$$
$$\leq \sum_{k=k_1}^{k_2-1} \max\left\{\frac{|\Delta w^k|}{w^k}, \frac{|\Delta w^k|}{w^{k+1}}\right\},$$

where $\Delta w^k = w^{k+1} - w^k$.

The first equality is trivial, and the second is an immediate consequence of the definition of Riemann integral. Now applying Lemma 1.1, since by Theorem 1.2 $|\Delta w^k| \leq \|z^{k+1} - z^k\| < 0.1$,

$$\frac{|\Delta w^k|}{w^{k+1}} \leq 1.1 \frac{|\Delta w^k|}{w^k}.$$

It follows that

$$|\log w^{k_1} - \log w^{k_2}| \leq 1.1 \sum_{k=k_1}^{k_2-1} \frac{|\Delta w^k|}{w^k}.$$

In particular, for $k_1 = k_i$ and $k_2 = k_{i+1}$, with $k_i, k_{i+1} \in T_j$, (1.34) gives

$$\mu \leq |\log w^{k_i} - \log w^{k_{i+1}}| \leq 1.1 \sum_{k=k_i}^{k_{i+1}-1} \frac{|\Delta w^k|}{w^k}.$$

Finally, the summation of the complete sequence (which is composed of $|T_j|$ partial summations as above) gives

$$|T_j| \mu \leq 1.1 \sum_{k=0}^{K} \frac{|\Delta w^k|}{w^k},$$

completing the proof.

Theorem 1.4. *Consider the sequence* (z^k) *generated by the algorithm and let* $T = \sum_{j=1}^{n} |T_j|$ *be the total number of rank-1 updates computed by the algorithm until iteration* K. *Then*

$$T \leq \sqrt{n}K.$$

PROOF. Applying Lemma 1.16,

$$T = \sum_{j=1}^{n} |T_j| \leq \frac{1.1}{\mu} \sum_{j=1}^{n} \sum_{k=0}^{K} \left| \frac{\Delta z_j^k}{z_j^k} \right|.$$

Inverting the order of the summations,

$$T \leq \frac{1.1}{\mu} \sum_{k=0}^{K} \left\| \left[\frac{\Delta z_j^k}{z_j^k} \right] \right\|_1.$$

But for any $y \in \mathbb{R}^n$, norms 1 and 2 are related by $\|y\|_1 \leq \sqrt{n}\|y\|$, and consequently

$$T \leq \frac{1.1}{\mu} \sqrt{n} \sum_{k=0}^{K} \|\Delta z^k\|_{z^k}.$$

Now using Theorem 1.2, for any $k = 0, \ldots, K$, $\|\Delta z^k\|_{z^k} \leq 0.04$ and it follows that

$$T \leq \frac{0.044}{\mu} \sqrt{n}K.$$

It is now enough to remember that $\mu > 0.09$ to complete the proof.

We can now prove the main complexity result.

Theorem 1.5. *Algorithm* 1.2 *with the updating procedure* 1.3 *solves the linear programming problem* (1.1) *in no more than* $O(n^3L)$ arithmetic operations.

PROOF. We showed in Theorem 1.3 that the algorithm terminates with an optimal solution in $K = O(\sqrt{n}L)$ iterations. Each iteration updates one projection matrix and computes a fixed number of matrix products.

The total number of operations needed per iteration excepting the projection matrix computation is then of the order $O(n^2)$, with a total figure of $O(n^{2.5}L)$ in K iterations.

The total number of operations needed for the projection updates is given by $T \times O(n^2)$, where T is the total number of rank-1 updates. By Theorem 5.2, $T \leq n^{0.5}K$, and consequently T is bounded by $O(nL)$. This gives a total figure of $O(n^3L)$, completing the proof.

The linear algebra computations are identical to the ones in Karmarkar's method [8] and can be carried with L bits of precision, as proved in [14]. This leads to a bound $O(n^3L^2)$ for the bit operations in our algorithm.

§6. Conclusion

Toward a Practical Algorithm

To develop a practical algorithm, the short steps must be made adaptive. This can easily be done by examining the N-R step computations:

$$h = -\frac{c_p}{\varepsilon_{k+1}} + d \quad \text{where } d = \bar{P}Y^{-1}e.$$

To make this step adaptive, two things should be done:

(i) Use a line search along direction h instead of a fixed step; that is, solve

$$\min_{\alpha>0} \{f_{\varepsilon_{k+1}(y+\alpha h)} | y + \alpha h \in \bar{S}\}.$$

(ii) Instead of a fixed penalty multiplier, let $\varepsilon_{k+1} = \varepsilon_k - \sigma$ and make σ vary in a "trust region" Γ.

Putting together these two procedures, we end with a bidirectional search

$$\min_{\alpha,\sigma>0} \left\{ f_{\varepsilon_k-\sigma}\left(y - \frac{\alpha c_p}{\varepsilon_k - \sigma} + \alpha d \right) \middle| \, y + \alpha h \in \bar{S}, \sigma \in \Gamma \right\}.$$

The region Γ must be such as to guarantee that the quadratic approximations are good. Several criteria may be used, and the best seems to be the following: a value of σ is acceptable if the search in α for this σ results in a value near 1.

Euler's Method

Since our method is a predictor-corrector algorithm, Euler's method immediately rings a bell. Instead of the simple elevator step (update of ε), we may follow a tangent to the path of optimizers. It is easy to see from Section 3.2 that a tangent direction for $x = e$ is given by

$$\frac{dx}{d\varepsilon} = \frac{1}{\varepsilon} e_p.$$

At point y,

$$\frac{dy}{d\varepsilon} = \frac{1}{\varepsilon_k} d$$

and associating with each value of $\Delta\varepsilon = -\sigma$

$$y_\sigma = y - \frac{\sigma}{\varepsilon_k} d,$$

the bidirectional search becomes

$$\min_{\alpha,\sigma>0} \{f_{\varepsilon_k-\sigma}(y_\sigma + \alpha h_\sigma) | \, y_\sigma + \alpha h_\sigma \in \bar{S}, \sigma \in \Gamma\},$$

where h_σ can be computed as before, $h_\sigma = h$, or by a new projection (with the same projection matrix):

$$h_\sigma = -\frac{c_p}{\varepsilon_k - \sigma} + d_\sigma, \quad \text{where } d_\sigma = \bar{P}Y_\sigma^{-1}e.$$

Notice that Euler's method does not introduce any new direction if h is kept unchanged, and no improvement in the result of a bidirectional search can be expected. The only modification to the basic bidirectional algorithm suggested by this procedure would be to recalculate d whenever the search procedure generates points far from the initial one.

This last observation falls into the idea developed in [6]: if the projection matrix is available at low computational cost, it is always advisable to perform steepest descent searches while they lead to good improvements in the function to be minimized. Summing up these observations, the best path toward a practical algorithm is in our opinion (and this must still be verified) the bidirectional search proposed above with recalculations of the vector d.

This bidirectional search is a variant of the general pattern followed by all conical projection algorithms and will be the subject of a forthcoming paper [7].

All these new algorithms boil down to a clever rearrangement of old techniques. Karmarkar showed that linear programming is indeed a particular case of nonlinear programming and that nice results can be obtained by specializing its methods to linear functions. Nothing is new, and all is new—like the fresh meats and garden vegetables rethought by the French nouvelle cuisine.

Acknowledgment

I would like to thank Prof. E. Polak, my host in Berkeley, for his advice and encouragement.

References

[1] K. Bayer and J. C. Lagarias, The non-linear geometry of linear programming, I. Affine and projective scaling trajectories, II. Legendre transform coordinates, III. Central trajectories, *Trans. Amer. Math. Soc.* (to appear).

[2] A. Fiacco and G. McCormick. *Nonlinear Programming: Sequential Unconstrained Minimization Techniques*, Wiley, New York, 1955.

[3] K. R. Frisch. The logarithmic potential method of convex programming, Memorandum, University Institute of Economics, Oslo, Norway (May 1955).

[4] C. B. Garcia and W. I. Zangwill, *Pathways to Solutions, Fixed Points, and Equilibria*, Prentice-Hall, Englewood cliffs, N.J., 1981.

[5] P. Gill, W. Murray, M. Saunders, J. Tomlin, and M. Wright, On projected Newton barrier methods for linear programming and an equivalence to Karmarkar's projective method, Report SOL 85-11, Systems Optimization

Laboratory, Department of Operations Research, Stanford University, Stanford, Calif. (1985).

[6] C. Gonzaga, A conical projection algorithm for linear programming, Memorandum UCB/ERL M85/61, Electronics Research Laboratory, University of California, Berkeley (July 1985).

[7] C. Gonzaga, Search directions for interior linear programming methods, Memorandum UCB/ERL M87/44, Electronics Research Laboratory, University of California, Berkeley (March 1987).

[8] N. Karmarkar, A new polynomial time algorithm for linear programming, *Combinatorica* **4** (1984), 373–395.

[9] F. A. Lootsma, *Numerical Methods for Nonlinear Optimization*, Academic Press, New York, 1972.

[10] N. Megiddo, Pathways to the optimal set in linear programming, Chapter 8, this volume.

[11] E. Polak, *Computational Method in Optimization*, Academic Press, New York, 1971.

[12] James Renegar, A polynomial-time algorithm based on Newton's method for linear programming. *Mathematical Programmings* **40** (1988), 50–94.

[13] M. Todd and B. Burrell, An extension of Karmarkar's algorithm for linear programming using dual variables, *Algorithmica* **1** (1986), 409–424.

[14] Pravin M. Vaidya, An algorithm for linear programming which requires $O(((m + n)n^2 + (m + n)^{1.5}n)L)$ arithmetic operations, *Proc. 19th Annual ACM Symposium on Theory of Computing* (1987), pp. 29–38.

CHAPTER 2

A Primal-Dual Interior Point Algorithm for Linear Programming

Masakazu Kojima, Shinji Mizuno, and Akiko Yoshise

Abstract. This chapter presents an algorithm that works simultaneously on primal and dual linear programming problems and generates a sequence of pairs of their interior feasible solutions. Along the sequence generated, the duality gap converges to zero at least linearly with a global convergence ratio $(1 - \eta/n)$; each iteration reduces the duality gap by at least η/n. Here n denotes the size of the problems and η a positive number depending on initial interior feasible solutions of the problems. The algorithm is based on an application of the classical logarithmic barrier function method to primal and dual linear programs, which has recently been proposed and studied by Megiddo.

§1. Introduction

This chapter deals with the pair of the standard form linear program and its dual.

(P) Minimize $c^T x$
subject to $x \in S = \{x \in R^n : Ax = b, x \geqq 0\}$.

(D) Maximize $b^T y$
subject to $(y, z) \in T = \{(y, z) \in R^{m+n} : A^T y + z = c, z \geqq 0\}$.

Here R^n denotes the n-dimensional Euclidean space, $c \in R^n$ and $b \in R^m$ are constant column vectors, and A is an $m \times n$ constant matrix. Throughout the chapter we impose the following assumptions on these problems:

(a) The set $S^I = \{x \in S : x > 0\}$ of the strictly positive feasible solutions of the primal problem (P) is nonempty.

(b) The set $T^I = \{(y, z) \in T : z > 0\}$ of the interior points of the feasible region T of the dual problem (D) is nonempty.
(c) Rank $A = m$.

By the well-known duality theorem, we know that both of the problems (P) and (D) have optimal solutions with a common value, say s^*. Furthermore, it is easily seen that the sets of the optimal solutions of (P) and (D) are bounded.

To solve the primal problem (P), we consider the problem of minimizing the objective function $c^T x$ with the additional logarithmic barrier term $-\mu \sum_{j=1}^{n} \log x_j$ over the relative interior S^I of the feasible region S:

$$(\mathrm{P}_\mu) \qquad \begin{array}{l} \text{Minimize } c^T x - \mu \sum_{j=1}^{n} \log x_j \\ \text{subject to } x \in S^I. \end{array}$$

Here μ is a penalty parameter. The logarithmic barrier function method was used for linear programs by Frish [2] and recently by Gill et al. [4] to develop a projected Newton barrier method in connection with Karmarkar's projective rescaling method [8, 9]. Megiddo [14] has made a deep analysis on the curve consisting of the solutions $x(\mu)$ of (P_μ) $(\mu > 0)$. He has shown some nice duality relations on the logarithmic barrier function method for a symmetric pair of primal and dual linear programs. A new algorithm we propose here will be founded on his analysis and duality relations. But we will deal with the nonsymmetric pair of the primal and dual problems, (P) and (D). So we will briefly explain them in terms of the nonsymmetric pair.

We first note that the objective function of (P_μ) is convex. Hence each problem (P_μ) has at most one local minimal solution. Furthermore, we can prove under assumptions (a) and (b) that each subproblem (P_μ) has a unique global minimal solution x (see Section 2 of Megiddo [14]), which is completely characterized by the Karush-Kuhn-Tucker stationary condition (see, for example, Mangasarian [13]):

$$\begin{aligned} c - \mu X^{-1} e - A^T y &= 0, \\ Ax - b = 0, \qquad x &> 0. \end{aligned}$$

Here $X = \mathrm{diag}(x_1, \ldots, x_n)$ denotes the diagonal matrix with the diagonal elements $x_1, \ldots, x_n$ and $e \in R^n$ the vector of 1's. Introducing a slack variable vector $z = \mu X^{-1} e$, we will write the stationary condition as

$$\begin{aligned} Zx - \mu e &= 0, \\ Ax - b = 0, \qquad x &> 0, \\ A^T y + z - c &= 0, \end{aligned} \tag{2.1}$$

where $Z = \mathrm{diag}(z_1, \ldots, z_n)$. The first equality can be further written component wise as

$$x_i z_i = \mu \qquad (i = 1, 2, \ldots, n).$$

That is, all the products of the complementary pair of the variables x_i and z_i $(i = 1, 2, \ldots, n)$ take a common value μ, the value of the penalty parameter.

Let $(x(\mu), y(\mu), z(\mu))$ denote a unique solution of the system (2.1) for each $\mu > 0$. Then each point $(y(\mu), z(\mu))$ lies in the interior of the dual feasible region T, that is, $(y(\mu), z(\mu)) \in T^I$. The gap between the primal objective value at $x(\mu) \in S^I$ and the dual objective value at $(y(\mu), z(\mu)) \in T^I$ turns out to be

$$c^T x(\mu) - b^T y(\mu) = (c^T - y(\mu)^T A) x(\mu) = z(\mu)^T x(\mu)$$
$$= e^T Z(\mu) x(\mu) = n\mu.$$

System (2.1) also gives a necessary and sufficient condition (the Karush-Kuhn-Tucker stationary condition) for $(y(\mu), z(\mu))$ to be a maximal solution of the subproblem:

$$(\mathrm{D}_\mu) \quad \begin{array}{l} \text{Maximize } b^T y + \mu \sum_{j=1}^{n} \log z_j \\ \text{subject to } (y, z) \in T^I. \end{array}$$

This subproblem can be regarded as an application of the barrier function method to the dual problem (D). See Theorem 3.1 of [14] for more detailed relations on the subproblems (P_μ) and (D_μ).

Let Γ denote the curve consisting of the solutions of the system (2.1), that is,

$$\Gamma = \{(x(\mu), y(\mu), z(\mu)) : \mu > 0\}.$$

As we take the limit $\mu \to 0$, the curve Γ leads to a pair of optimal solutions x^* of problem (P) and (y^*, z^*) of problem (D), which satisfies the strong complementarity such that $z_i^* = 0$ if and only if $x_i^* > 0$ (Proposition 3.2 of [14]). We have observed that if $(x(\mu), y(\mu), z(\mu))$ is a solution of the system (2.1) or if $(x(\mu), y(\mu), z(\mu))$ lies on the curve Γ then

$$x(\mu) \in S^I, \tag{2.2}$$

$$(y(\mu), z(\mu)) \in T^I, \tag{2.3}$$

and

$$c^T x(\mu) - b^T y(\mu) = n\mu. \tag{2.4}$$

Thus, tracing the curve Γ, which has been proposed in [14], serves as a theoretical model for a class of primal-dual interior point algorithms for linear programs. The new algorithm, which will be described in Section 2, can be regarded as a practical implementation of this model using Newton's method. Newton's method (or projected Newton's method) has been used in several interior point algorithms for linear programs (Gill et al. [4], Imai [6], Iri and Imai [7], Renegar [16], Yamashita [21]). In those algorithms it is applied to minimization of barrier functions or penalty functions, whereas in our algorithm it will be applied to the system of equations (2.1) that has been derived as the Karush-Kuhn-Tucker stationary condition for problem P_μ, the minimization of the logarithmic barrier function over the interior of the primal feasible region.

Given an initial point $(x^0, y^0, z^0) \in S^I \times T^I$, the algorithm generates a sequence $\{(x^k, y^k, z^k)\}$ in $S^I \times T^I$. To measure a "deviation" from the curve Γ at each $(x^k, y^k, z^k) \in S^I \times T^I$, we introduce the quantity

$$\pi^k = f^k_{\text{ave}}/f^k_{\text{min}}, \tag{2.5}$$

where

$$\begin{aligned} f_i^k &= x_i^k z_i^k \qquad (i = 1, 2, \ldots, n), \\ f^k_{\text{min}} &= \min\{f_i^k : i = 1, 2, \ldots, n\}, \\ f^k_{\text{ave}} &= \left\{\left(\sum_{i=1}^n f_i^k\right)\Big/n\right\} \qquad \text{(the average of } f_i^k\text{'s).} \end{aligned} \tag{2.6}$$

Obviously, we see that $\pi^k \geqq 1$ and that $(x^k, y^k, z^k) \in S^I \times T^I$ lies on the curve Γ if and only if $\pi^k = 1$. When the deviation π^0 at the initial point $(x^0, y^0, z^0) \in S^I \times T^I$ is large, we reduce not only the duality gap but also the deviation. Keeping a small deviation, we then direct our main effort to reducing the duality gap.

Several properties of the global convergence of the algorithm will be investigated in Section 3. The algorithm has two control parameters. Specifically, we will see that if we make a suitable choice of the parameters then the generated sequence $\{(x^k, y^k, z^k) \in S^I \times T^I\}$ satisfies the inequalities

$$\begin{aligned} c^T x^{k+1} - b^T y^{k+1} &\leqq (1 - 2/(n\pi^k))(c^T x^k - b^T y^k), \\ \pi^{k+1} - 2 &\leqq (1 - 1/(n+1))(\pi^k - 2) \quad \text{if } \pi^k > 2, \\ \pi^{k+1} &\leqq 2 \quad \text{if } \pi^k \leqq 2. \end{aligned}$$

(Corollary 2.2). The first inequality ensures that the duality gap decreases monotonically. By the second, we see that the deviation π^k gets smaller than 3 within at most $O(n \log \pi^0)$ iterations, and then the duality gap converges to zero linearly with the convergence rate at most $(1 - 2/(3n))$. Each iteration requires $O(n^3)$ arithmetic operations.

Generally an initial primal and dual feasible solution $(x^0, y^0, z^0) \in S^I \times T^I$ is not known, so we have to incorporate a phase 1 procedure. This will be explained in Section 4, where we provide artificial primal and dual linear programs with a known feasible solution.

System (2.1) gives some significant insights into the interior point algorithms developed so far (e.g., Adler et al. [1], Gay [3], Gill et al. [4], Imai [6], Iri and Imai [7], Karmarkar [8,9], Renegar [16], Todd and Burrell [19], Yamashita [21], Ye and Kojima [22]). Originally these algorithms were designed to apply to various different forms of linear programs. But when they are adapted to the standard form linear program (P), most of them have the common property that the search direction along which new interior feasible solutions are generated from a current feasible solution x is composed of the two directions, $XPXc$ and XPe, although the coefficients associated with these two directions vary in different algorithms. Here P denotes the orthogonal projection matrix onto the subspace $\{w \in R^n : AXw = 0\}$, $X = \mathrm{diag}(x_1, x_2, \ldots, x_n)$ an $n \times n$ diagonal matrix, and e the n-dimensional vector of 1's. This fact has recently been pointed out by Yamashita [21]. If (x, y, z) lies on the curve Γ consisting of the solutions of system (2.1), then these two directions coincide with the tangent vector of the curve Γ. This indicates that the local behaviors of these algorithms and ours are quite similar in neighborhoods of the curve Γ. We refer to the paper [15] by Megiddo and Shub, who have studied the boundary behavior of Karmarkar's projective rescaling algorithm and the linear rescaling algorithm.

In the discussions above, the matrix X serves as a linear (or affine) rescaling that transforms the current feasible solution x to the vector e in the primal space. When we deal with the dual problem (D), the diagonal matrix $Z^{-1} = \mathrm{diag}(1/z_1, 1/z_2, \ldots, 1/z_n)$ plays this role [1]. It is also interesting to note that the rescaling matrix X for the primal problem (P) has been utilized to generate feasible solutions of the dual problem (D) to get lower bounds for the primal objective function in some of the algorithms listed above [3, 19, 22]. By the first equality of the system (2.1) we see that if $(x, y, z) \in \Gamma$, then the rescalings by the matrix X in the primal space and by the matrix Z^{-1} in the dual space differ only by a constant multiple so that they are essentially the same. In our algorithm the common rescaling matrix $(XZ^{-1})^{1/2}$, which is the geometric mean of the primal rescaling matrix X and the dual rescaling matrix Z^{-1}, will be applied simultaneously to the primal and dual spaces even if the point (x, y, z) does not lie on the curve Γ.

§2. Algorithm

We begin by deriving the Newton direction of the system of equations (2.1) at $(x^k, y^k, z^k) \in S^I \times T^I$. If we denote the right-hand side of equations (2.1) by $H(x, y, z)$, then the Newton direction $(\Delta x, \Delta y, \Delta z)$ at (x^k, y^k, z^k) is defined by the system of linear equations

$$H(x^k, y^k, z^k) - D_x H \Delta x - D_y H \Delta y - D_z H \Delta z = 0,$$

where $D_x H$, $D_y H$, and $D_z H$ denote the Jacobian matrices of the mapping H at (x^k, y^k, z^k) with respect to the variable vectors x, y, and z, respectively. It follows that

$$\begin{aligned} Z\Delta x + X\Delta z &= v(\mu), \\ A\Delta x &= 0, \\ A^T \Delta y + \Delta z &= 0, \end{aligned} \tag{2.7}$$

where

$$X = \mathrm{diag}(x_1^k, x_2^k, \ldots, x_n^k),$$
$$Z = \mathrm{diag}(z_1^k, z_2^k, \ldots, z_n^k),$$
$$v(\mu) = Xz^k - \mu e \in R^n.$$

It should be noted that a solution of this system depends linearly on the penalty parameter μ, so we will write a solution as $(\Delta x(\mu), \Delta y(\mu), \Delta z(\mu))$. By a simple calculation, we obtain

$$\Delta x(\mu) = \{Z^{-1} - Z^{-1}XA^T(AZ^{-1}XA^T)^{-1}AZ^{-1}\}v(\mu),$$
$$\Delta y(\mu) = -(AZ^{-1}XA^T)^{-1}AZ^{-1}v(\mu),$$
$$\Delta z(\mu) = A^T(AZ^{-1}XA^T)^{-1}AZ^{-1}v(\mu).$$

We now introduce the diagonal matrix

$$D = (XZ^{-1})^{1/2} = \mathrm{diag}((x_1^k/z_1^k)^{1/2}, \ldots, (x_n^k/z_n^k)^{1/2}),$$

the vector-valued linear function

$$u(\mu) = (XZ)^{1/2}e - \mu(XZ)^{-1/2}e \in R^n \tag{2.8}$$

or

$$u_i(\mu) = (x_i^k z_i^k)^{1/2} - \mu(x_i^k z_i^k)^{-1/2} \qquad (i = 1, 2, \ldots, n)$$

in the variable μ, and the orthogonal projection matrix

$$Q = DA^T(AD^2A^T)^{-1}AD$$

onto the subspace $\{DA^Ty : y \in R^m\}$. Then the Newton direction $(\Delta x(\mu), \Delta y(\mu), \Delta z(\mu))$ can be written as

$$\begin{aligned} \Delta x(\mu) &= D(I - Q)u(\mu), \\ \Delta y(\mu) &= -(AD^2A^T)^{-1}ADu(\mu), \\ \Delta z(\mu) &= D^{-1}Qu(\mu). \end{aligned} \tag{2.9}$$

These equalities and (2.7) imply

$$\begin{aligned} D^{-1}\Delta x(\mu) + D\Delta z(\mu) &= u(\mu), \\ 0 = \Delta x(\mu)^T\Delta z(\mu) &= (D^{-1}\Delta x(\mu))^T(D\Delta z(\mu)). \end{aligned} \tag{2.10}$$

At the start of the algorithm described below, we fix two real parameters σ and τ such that $0 \leqq \tau < \sigma < 1$, which control the penalty parameter μ and the step length, respectively. Define π^k, f_i^k $(i = 1, 2, \ldots, n)$, f_{ave}^k, and $f_{\min}^k$ by (2.5) and (2.6). We will set the penalty parameter $\mu = \sigma f_{\mathrm{ave}}^k$ and then compute a new point, the $(k+1)$th point from (x^k, y^k, z^k) in the direction $-(\Delta x(\mu), \Delta y(\mu), \Delta z(\mu))$. That is, $(x^{k+1}, y^{k+1}, z^{k+1})$ can be written in the form

$$
\begin{aligned}
x(\alpha, \mu) &= x^k - \alpha \Delta x(\mu), \\
y(\alpha, \mu) &= y^k - \alpha \Delta y(\mu), \\
z(\alpha, \mu) &= z^k - \alpha \Delta z(\mu).
\end{aligned} \tag{2.11}
$$

Here α denotes a step length.

As the step length α changes, the product $f_i = x_i z_i$ of each complementary pair of variables x_i and z_i changes quadratically:

$$
\begin{aligned}
f_i(\alpha, \mu) &= x_i(\alpha, \mu) z_i(\alpha, \mu) \\
&= x_i^k z_i^k - \alpha(x_i^k \Delta z_i(\mu) + z_i^k \Delta x_i(\mu)) + \Delta x_i(\mu) \Delta z_i(\mu) \alpha^2 \\
&= f_i^k - (f_i^k - \mu)\alpha + \Delta x_i(\mu) \Delta z_i(\mu) \alpha^2 \qquad (i = 1, 2, \ldots, n),
\end{aligned} \tag{2.12}
$$

(since $x_i^k \Delta z_i(\mu) + z_i^k \Delta x_i(\mu) = f_i^k - \mu$ by (2.7)) and the average f_{ave} of f_i $(i = 1, 2, \ldots, n)$ as well as the duality gap $c^T x - b^T y$ changes linearly:

$$
\begin{aligned}
f_{\text{ave}}(\alpha, \mu) &= \left\{ \sum_{i=1}^{n} f_i(\alpha, \mu) \right\} / n \\
&= x(\alpha, \mu)^T z(\alpha, \mu)/n \\
&= f_{\text{ave}}^k - (f_{\text{ave}}^k - \mu)\alpha \quad (\text{since } \Delta x(\mu)^T \Delta z(\mu) = 0 \text{ by (2.10)})
\end{aligned} \tag{2.13}
$$

$$
\begin{aligned}
c^T x(\alpha, \mu) - b^T y(\alpha, \mu) &= x(\alpha, \mu)^T z(\alpha, \mu) \\
&= x^{kT} z^k - (x^{kT} z^k - n\mu)\alpha \\
&= (c^T x^k - b^T y^k) - \{(c^T x^k - b^T y^k) - n\mu\}\alpha.
\end{aligned} \tag{2.14}
$$

Figure 2.1 illustrates the functions f_i and f_{ave}. If we neglected the quadratic term $\Delta x_i(\mu) \Delta z_i(\mu) \alpha^2$ in the function f_i, it would induce a line through the two points $(0, f_i^k)$ and $(1, \mu)$. It should be noted that the latter point does not depend on the coordinate i.

The step length α^k will be determined by

$$
\begin{aligned}
\beta^k &= \max\{\alpha \colon f_i(\alpha', \mu) \geqq \psi(\alpha') \quad \text{for all } \alpha' \in (0, \alpha) \text{ and } i = 1, 2, \ldots, n\}, \\
\alpha^k &= \min\{1, \beta^k\},
\end{aligned} \tag{2.15}
$$

where

$$
\psi(\alpha) = f_{\min}^k - (f_{\min}^k - \tau f_{\text{ave}}^k)\alpha. \tag{2.16}
$$

See Figure 2.2.

Now we are ready to state the algorithm.

Algorithm

Step 0: Let $(x^0, y^0, z^0) \in S^I \times T^I$ be an initial point and ε a tolerance for duality gap. Let σ and τ be real numbers such that $0 \leqq \tau < \sigma < 1$. Let $k = 0$.

Step 1: $c^T x^k - b^T y^k < \varepsilon$ then stop.

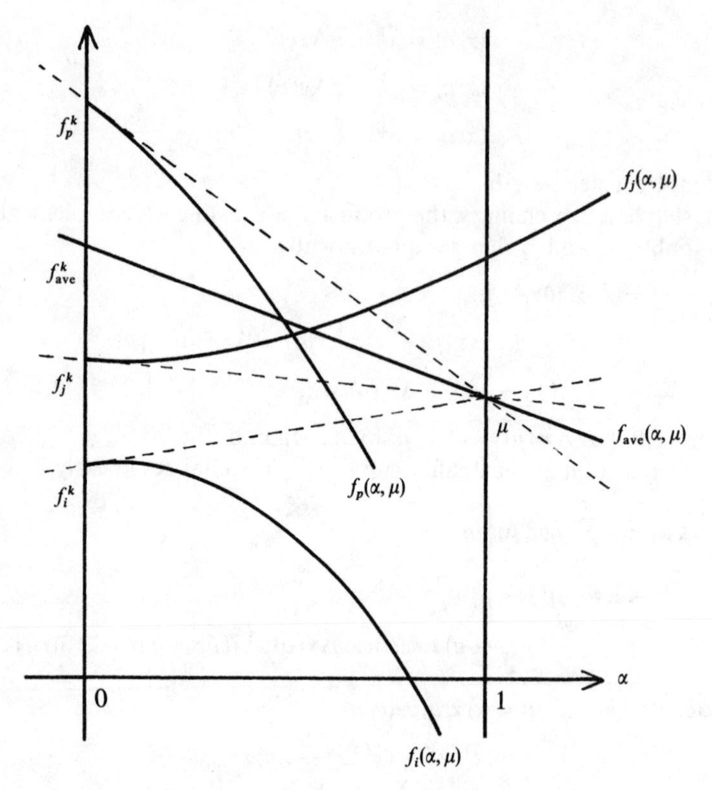

Figure 2.1. The function f_i $(i = 1, 2, \ldots, n)$ and f_{ave}.

Step 2: Define f_{ave}^k and f_{min}^k by (2.6).
Step 3: Let $\mu = \sigma f_{ave}^k$ and compute $(\Delta x(\mu), \Delta y(\mu), \Delta z(\mu))$ by (2.9).
Step 4: Compute β^k and α^k by (2.15).
Step 5: Let

$$\begin{aligned} x^{k+1} &= x(\alpha^k, \mu), \\ y^{k+1} &= y(\alpha^k, \mu), \\ z^{k+1} &= z(\alpha^k, \mu), \end{aligned} \tag{2.17}$$

where $x(\alpha, \mu)$, $y(\alpha,\mu)$, and $z(\alpha,\mu)$ are given by (2.11).
Step 6: Let $k = k + 1$. Go to step 1.

§3. Global Convergence Properties

In this section we show some global convergence properties of the algorithm. The following lemma provides a common lower bound for the functions f_i $(i = 1, 2, \ldots, n)$ defined by (2.12).

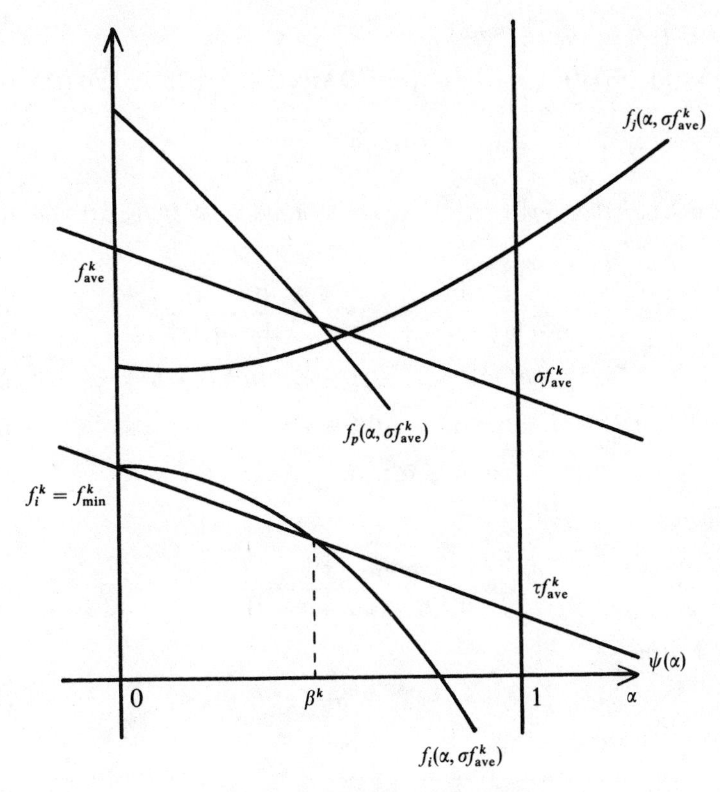

Figure 2.2. Computation of β^k.

Lemma. *Let*

$$\zeta(\alpha, \mu) = f^k_{\min} - (f^k_{\min} - \mu)\alpha - \|u(\mu)\|^2\alpha^2/4.$$

Then

$$f_i(\alpha, \mu) \geqq \zeta(\alpha, \mu) \quad \text{for any } \alpha \in (0, 1)\ (i = 1, 2, \ldots, n).$$

Proof. By (2.12) we have

$$f_i(\alpha, \mu) = f^k_i - (f^k_i - \mu)\alpha + (D^{-1}\Delta x(\mu))_i(D\Delta z(\mu))_i\alpha^2.$$

The first two constant and linear terms on the right-hand side are bounded by $f^k_{\min} - (f^k_{\min} - \mu)\alpha$ for all $\alpha \in [0, 1]$. So we have only to evaluate the last quadratic term. The vectors $D^{-1}\Delta x(\mu)$ and $D\Delta z(\mu)$ satisfy (2.10). In other words, the vector $u(\mu)$ is the orthogonal sum of these two vectors. Hence

$$\begin{aligned}(D^{-1}\Delta x(\mu))_j(D\Delta z(\mu))_j &= (u(\mu)_j - (D\Delta z(\mu))_j)(D\Delta z(\mu))_j \\ &\leqq |u(\mu)_j|^2/4\end{aligned}$$

for all $j = 1, 2, \ldots, n$. Therefore

$$(D^{-1}\Delta x(\mu))_i(D\Delta z(\mu))_i = (D^{-1}\Delta x(\mu))^T(D\Delta z(\mu)) - \sum_{j \neq i} (D^{-1}\Delta x(\mu))_j(D\Delta z(\mu))_j$$
$$\geqq 0 - \|u(\mu)\|^2/4.$$

Theorem 2.1 *If the step length α^k determined at step 4 of the* k*th iteration is less than 1, then*

$$\alpha^k \geqq \frac{4(\sigma - \tau)}{n(1 - 2\sigma + \pi^k\sigma^2)} \geqq \frac{4(\sigma - \tau)}{n(1 + \sigma^2)\pi^k}, \tag{2.18}$$

$$c^T x^{k+1} - b^T y^{k+1} = (1 - (1 - \sigma)\alpha^k)(c^T x^k - b^T y^k), \tag{2.19}$$

$$\pi^{k+1} - \sigma/\tau \leqq (1 - \nu)(\pi^k - \sigma/\tau) \quad \text{if } \sigma/\tau < \pi^k, \tag{2.20}$$

$$\pi^{k+1} \leqq \sigma/\tau \quad \text{if } \pi^k \leqq \sigma/\tau \tag{2.21}$$

hold, where

$$\nu = \frac{4(\sigma - \tau)\tau}{n(1 + \sigma^2) + 4(\sigma - \tau)\tau}. \tag{2.22}$$

If $\alpha^k = 1$ then

$$c^T x^{k+1} - b^T y^{k+1} = \sigma(c^T x^k - b^T y^k) \tag{2.23}$$

and

$$\pi^{k+1} \leqq \sigma/\tau \tag{2.24}$$

hold.

PROOF. We first consider the case $\alpha^k < 1$. Let

$$\gamma = \max\{\alpha: \zeta(\alpha, \sigma f^k_{\text{ave}}) \geqq \psi(\alpha)\}.$$

Since α^k is determined by (2.15) with $\mu = \sigma f^k_{\text{ave}}$, we see by the lemma that $\alpha^k = \beta^k \geqq \gamma$. By computing a positive solution of the equation

$$\zeta(\alpha, \sigma f^k_{\text{ave}}) = \psi(\alpha),$$

that is,

$$f^k_{\min} - (f^k_{\min} - \sigma f^k_{\text{ave}})\alpha - \|u(\sigma f^k_{\text{ave}})\|^2\alpha^2/4$$
$$= f^k_{\min} - (f^k_{\min} - \tau f^k_{\text{ave}})\alpha,$$

we obtain

$$\alpha^k = \beta^k \geqq \gamma = 4(\sigma - \tau)f^k_{\text{ave}}/\|u(\sigma f^k_{\text{ave}})\|^2. \tag{2.25}$$

On the other hand, by the definition (2.8) of $u(\mu)$, we see

$$
\begin{aligned}
\|u(\sigma f_{\text{ave}}^k)\|^2 &= \sum_{i=1}^n \{(f_i^k)^{1/2} - \sigma f_{\text{ave}}^k (f_i^k)^{-1/2}\}^2 \\
&= n f_{\text{ave}}^k - 2\sigma n f_{\text{ave}}^k + \sigma^2 (f_{\text{ave}}^k)^2 \sum_{i=1}^n (f_i^k)^{-1} \\
&\leqq n f_{\text{ave}}^k (1 - 2\sigma + \sigma^2 f_{\text{ave}}^k / f_{\min}^k) \\
&= n f_{\text{ave}}^k (1 - 2\sigma + \sigma^2 \pi^k) \\
&\leqq n f_{\text{ave}}^k (1 + \sigma^2) \pi^k \quad (\text{since } \pi^k \geqq 1). \qquad (2.26)
\end{aligned}
$$

Substituting the right-hand sides of the last two inequalities above into (2.25), we obtain inequality (2.18).

Inequality (2.19) follows from the construction (2.17) of $(x^{k+1}, y^{k+1}, z^{k+1})$, $\mu = \sigma f_{\text{ave}}^k = \sigma(c^T x^k - b^T y^k)/n$, and (2.14). Also see (2.11) for the definition of the mappings x, y, and z.

Now we show inequalities (2.20) and (2.21). By the definition (2.15) of the step length α^k and the construction of $f_{\min}^{k+1}$, we have

$$
\begin{aligned}
f_{\min}^{k+1} &= \min\{f_i(\alpha^k) : i = 1, 2, \ldots, n\} \\
&= \psi(\alpha^k) \\
&= f_{\min}^k - (f_{\min} - \tau f_{\text{ave}}^k)\alpha^k,
\end{aligned}
$$

By (2.13) we also see

$$
\begin{aligned}
f_{\text{ave}}^{k+1} &= f_{\text{ave}}(\alpha^k, \sigma f_{\text{ave}}^k) \\
&= f_{\text{ave}}^k - (f_{\text{ave}}^k - \sigma f_{\text{ave}}^k)\alpha^k.
\end{aligned}
$$

Hence

$$
\begin{aligned}
\pi^{k+1} - \sigma/\tau &= \frac{f_{\text{ave}}^k - (f_{\text{ave}}^k - \sigma f_{\text{ave}}^k)\alpha^k}{f_{\min}^k - (f_{\min}^k - \tau f_{\text{ave}}^k)\alpha^k} - \frac{\sigma}{\tau} \\
&= \frac{(1 - \alpha^k)(f_{\text{ave}}^k \tau - f_{\min}^k \sigma)}{\{f_{\min}^k - (f_{\min}^k - \tau f_{\text{ave}}^k)\alpha^k\}\tau} \\
&= \left\{1 - \frac{\tau \pi^k \alpha^k}{1 - (1 - \tau \pi^k)\alpha^k}\right\}(\pi^k - \sigma/\tau). \qquad (2.27)
\end{aligned}
$$

If $\pi^k \leqq \sigma/\tau$, then the right-hand side of the last equality is nonpositive; hence so is the left-hand side. This proves (2.21). Now assume that $\sigma/\tau < \pi^k$. Then the above equality implies

$$
\pi^{k+1} - \sigma/\tau \leqq \left\{1 - \frac{\tau \pi^k \alpha^k}{1 + \tau \pi^k \alpha^k}\right\}(\pi^k - \sigma/\tau). \qquad (2.28)
$$

On the other hand, we see from inequality (2.18) that

$$
\pi^k \alpha^k \geqq 4(\sigma - \tau)/\{n(1 + \sigma^2)\}.
$$

By substituting this inequality into the right-hand side of (2.28), we obtain (2.20).

Finally we deal with the case $\alpha^k = 1$ and show (2.23) and (2.24). Equality (2.23) can be derived in the same way as in the case of (2.19). Since $f_i(1, \sigma f_{\text{ave}}^k) \geqq \psi(1)$ for all $i = 1, 2, \ldots, n$, we obtain

$$\begin{aligned} \pi^{k+1} &= f_{\text{ave}}^{k+1} / f_{\min}^{k+1} \\ &\leqq f_{\text{ave}}(1, \sigma f_{\text{ave}}^k)/\psi(1) \\ &= \sigma/\tau. \end{aligned}$$

This completes the proof.

In view of the theorem above, if k^* is the earliest iteration at which $\pi^{k^*} \leqq \sigma/\tau$ then

$$\sigma/\tau < \pi^k \leqq (1 - v)^k(\pi^0 - \sigma/\tau) + \sigma/\tau \qquad (k = 0, 1, \ldots, k^* - 1)$$

and

$$\pi^k \leqq \sigma/\tau \qquad (k = k^*, k^* + 1, \ldots).$$

Here v is a positive constant given by (2.22). If such a k^* does not exist, that is, if $\pi^k > \sigma/\tau$ throughout all the iterations then

$$\sigma/\tau < \pi^k \leqq (1 - v)^k(\pi^0 - \sigma/\tau) + \sigma/\tau \qquad (k = 0, 1, \ldots).$$

In both cases we have that

$$\pi^k \leqq \max\{\sigma/\tau, \pi^0\} \qquad (k = 0, 1, 2, \ldots)$$

and that π^k gets smaller than $(\sigma/\tau) + 1$ in at most $O(n \log \pi^0)$ iterations, say $\hat{k}$ iterations.

Now assume that $k \geqq \hat{k}$. Then we have, by inequality (2.18),

$$(1 - \sigma)\alpha^k \geqq \frac{(1 - \sigma)4(\sigma - \tau)}{n(1 + \sigma^2)((\sigma/\tau) + 1)}.$$

By inequality (2.19), the duality gap $c^T x^k - b^T y^k$ attains the given accuracy ε and the iteration stops in at most $O(n \log((c^T x^0 - b^T y^0)/\varepsilon)$ additional iterations.

We summarize the above results in the following corollary.

Corollary 2.1. *The algorithm terminates in at most*

$$O(n \log \pi^0) + O(n \log((c^T x^0 - b^T y^0)/\varepsilon))$$

iterations.

In practical implementations of the algorithm, there may be various ways of setting the control parameters σ and τ such that $0 \leqq \tau < \sigma < 1$. Although the values of these parameters are fixed at the start of the iteration of the algorithm and the fixed values are used throughout the iterations, we could change those values at each iteration. As a special case, we obtain:

Corollary 2.2. *Suppose that* $\sigma = 1/2$ *and* $\tau = 1/4$. *Then*

$$\alpha^k \geqq 4/(n\pi^k),$$
$$c^T x^{k+1} - b^T y^{k+1} \leqq (1 - 2/(n\pi^k))(c^T x^k - b^T y^k),$$
$$\pi^{k+1} - 2 \leqq (1 - 1/(1 + n))(\pi^k - 2) \quad \text{if } 2 < \pi^k,$$
$$\pi^{k+1} \leqq 2 \quad \text{if } \pi^k \leqq 2.$$

PROOF. We have only to show the third inequality because the others follow immediately from Theorem 2.1. By the first inequality of the corollary, we see $\pi^k \alpha^k \geqq 4/n$. Hence, substituting this inequality, $\sigma = 1/2$, and $\tau = 1/4$ into (2.28), we obtain the desired result.

In view of the second inequality of Corollary 2.2, the kth iteration reduces the gap between the lower bound $c^T x^k$ and the upper bound $b^T y^k$ for a common optimal value of (P) and (D) by at least $2/(n\pi^k)$. The following theorem shows a method of getting better lower and upper bounds for the optimal value at each iteration; if we set $\mu = 0$ and take the largest step size α satisfying the primal and dual feasibility conditions $x(\alpha,0) \in S$ and $(y(\alpha, 0), z(\alpha, 0)) \in T$, then we can reduce the gap by at least $1/(n\pi^k)^{1/2}$.

Theorem 2.2. *Define*

$\hat{\theta} = \max\{\alpha: f_i(\alpha', 0) \geqq 0$ *for all* $\alpha' \in [0, \alpha]$ $(i = 1, 2, \ldots, n)\}$ *at step* 3 *of the algorithm. Let*

$$\hat{x} = x(\hat{\theta}, 0), \qquad \hat{y} = y(\hat{\theta}, 0), \qquad \hat{z} = z(\hat{\theta}, 0).$$

Then $\hat{x}$ *and* $(\hat{y}, \hat{z})$ *are feasible solutions of* (P) *and* (D), *respectively, and the inequality*

$$c^T \hat{x} - b^T \hat{y} \leqq (1 - 1/(n\pi^k)^{1/2})(c^T x^k - b^T y^k)$$

holds.

PROOF. By the construction of the mappings x, y, and z whose definition has been given by (2.11), $\hat{x}$ and $(\hat{y}, \hat{z})$ satisfy the equality constraints $A\hat{x} = b$ and $A^T \hat{y} + \hat{z} = c$. Since $f_i(\alpha, 0) = x_i(\alpha, 0) z_i(\alpha, 0)$ for every $\alpha \geqq 0$ $(i = 1, 2, \ldots, n)$ by definition (2.12), the nonnegativity $\hat{x} \geqq 0$ and $\hat{z} \geqq 0$ follows from the definition of $\hat{\theta}$. Thus we have shown the first statement of the theorem. By inequality (2.14) with $\mu = 0$ and $\alpha = \hat{\theta}$, we have

$$0 \leqq c^T \hat{x} - b^T \hat{y} = (1 - \hat{\theta})(c^T x^k - b^T y^k). \tag{2.29}$$

This implies that $\hat{\theta} \leqq 1$. Applying the lemma as in the proof of Theorem 2.1, we then see that $\hat{\theta}$ is not less than a positive solution of the equation

$$\zeta(\alpha, 0) = f^k_{\min} - f^k_{\min}\alpha - \|u(0)\|^2\alpha^2/4 = 0.$$

It follows that

$$\begin{aligned}\theta &\geqq 2\{-f^k_{\min} + ((f^k_{\min})^2 + f^k_{\min}\|u(0)\|^2)^{1/2}\}/\|u(0)\|^2 \\ &= 2/\{(1 + \|u(0)\|^2/f^k_{\min})^{1/2} + 1\}.\end{aligned}$$

On the other hand, we see by definition (2.8) of $u(\mu)$ that

$$\|u(0)\|^2 = nf^k_{\mathrm{ave}} = nf^k_{\min}\pi^k.$$

Hence

$$\begin{aligned}\theta &\geqq 2/\{(1 + n\pi^k)^{1/2} + 1\} \\ &\geqq 1/(n\pi^k)^{1/2}.\end{aligned}$$

By substituting this inequality into (2.29), we obtain the desired inequality.

Remark. If (x^k, y^k, z^k) lies on the curves Γ consisting of the solutions of the system (2.1) then $\pi^k = 1$. In this case, the Newton direction $(\Delta x(0), \Delta y(0), \Delta z(0))$ defined by (2.9) with $\mu = 0$ coincides with the tangent vector of the curve Γ. Theorem 2.2 above ensures that the linear extrapolation of the curve Γ along the tangent vector to the boundary of the feasible regions in the primal and dual spaces generates a pair of primal and dual feasible solutions that reduces the interval containing the common optimal value of the problems (P) and (D) by at least $n^{-1/2}$.

§4. Artificial Primal and Dual Linear Programs for Initiating the Algorithm

In the discussions so far, we have been assuming that a feasible solution $x^0 \in S^0$ of the primal problem (P) and a feasible solution $(y^0, z^0) \in T^0$ of the dual problem (D) are available so that we can immediately start the algorithm. In general, however, it is necessary to provide a means for determining an $(x, y, z) \in S^0 \times T^0$ so that the algorithm can be initiated. The method described below is essentially due to Megiddo [14]. See Section 3 of [14].

Let $(x^0, y^0, z^0) \in R^{n+m+n}$ be an arbitrary point satisfying $x^0 > 0$ and $z^0 > 0$. Let κ and λ be sufficiently large positive numbers, whose magnitude will be specified below. Corresponding to the pair of primal and dual linear programs, (P) and (D), which we want to solve, we consider the following pair of artificial primal and dual linear programs:

$$(\mathrm{P}')\quad \begin{array}{lll} \text{Minimize} & c^T x + \kappa x_{n+1} & \\ \text{subject to} & Ax + (b - Ax^0)x_{n+1} & = b, \\ & (A^T y^0 + z^0 - c)^T x \qquad + x_{n+2} & = \lambda, \\ & (x, x_{n+1}, x_{n+2}) \geqq 0, & \end{array}$$

where x_{n+1} and x_{n+2} are artificial real variables.

$$
(\mathrm{D}')\quad
\begin{array}{lllr}
\text{Maximize} & b^T y + & \lambda y_{m+1} & \\
\text{subject to} & A^T y + (A^T y^0 + z^0 - c) y_{m+1} + z & & = c, \\
 & (b - Ax^0)^T y & + z_{n+1} & = \kappa, \\
 & & y_{m+1} + z_{n+2} & = 0, \\
 & (z, z_{n+1}, z_{n+2}) \geqq 0, & &
\end{array}
$$

where y_{m+1}, z_{n+1}, and z_{n+2} are artificial real variables.

We need to take κ and λ satisfying

$$\lambda > (A^T y^0 + z^0 - c)^T x^0 \tag{2.30}$$

and

$$\kappa > (b - Ax^0)^T y^0, \tag{2.31}$$

so that $(x^0, x_{n+1}^0, x_{n+2}^0)$ and $(y^0, y_{m+1}^0, z^0, z_{n+1}^0, z_{n+2}^0)$ are feasible solutions of (P′) and (D′), respectively, where

$$
\begin{aligned}
x_{n+1}^0 &= 1, \\
x_{n+2}^0 &= \lambda - (A^T y^0 + z^0 - c)^T x^0, \\
y_{m+1}^0 &= -1, \\
z_{n+1}^0 &= \kappa - (b - Ax^0)^T y^0, \\
z_{n+2}^0 &= 1.
\end{aligned}
$$

Therefore, we can apply the algorithm to the artificial problems (P′) and (D′) with these initial feasible solutions.

Let x^* and (y^*, z^*) be optimal solutions of the original problems (P) and (D), respectively. The theorem below gives a sufficient condition on the constants κ and λ for the algorithm applied to the artificial problems (P′) and (D′) to succeed in generating approximate solutions of the original problems (P) and (D).

Theorem 2.3. *In addition to* (2.30) *and* (2.31), *suppose that*

$$\lambda > (A^T y^0 + z^0 - c)^T x^* \tag{2.32}$$

and

$$\kappa > (b - Ax^0)^T y^*. \tag{2.33}$$

Then the following (a) *and* (b) *hold*:

(a) *A feasible solution* $(\hat{x}, \hat{x}_{n+1}, \hat{x}_{n+2})$ *of* (P′) *is minimal if and only if* $\hat{x}$ *is a minimal solution of* (P) *and* $\hat{x}_{n+1} = 0$.

(b) *A feasible solution* $(\hat{y}, \hat{y}_{m+1}, \hat{z}, \hat{z}_{n+1} \hat{z}_{n+2})$ *of* (D′) *is maximal if and only if* $(\hat{y}, \hat{z})$ *is a maximal solution of* (D) *and* $\hat{y}_{m+1} = 0$.

PROOF. Define

$$x_{n+1}^* = 0,$$

$$x_{n+2}^* = \lambda - (A^T y^0 + z^0 - c)^T x^*.$$

Then $(x^*, x_{n+1}^*, x_{n+2}^*)$ is a feasible solution. Let (x, x_{n+1}, x_{n+2}) be an arbitrary feasible solution of (P′) with $x_{n+1} > 0$. Then we see

$$\begin{aligned} c^T x^* + \kappa x_{n+1}^* &= y^{*T} b \\ &= y^{*T}\{Ax + (b - Ax^0)x_{n+1}\} \\ &< (c - z^*)^T x + \kappa x_{n+1} \quad (\text{by } Ay^* + z^* = c,\ x_{n+1} > 0, \text{ and } (2.33)) \\ &\leqq c^T x + \kappa x_{n+1} \quad (\text{by } (z^*)^T x \leqq 0). \end{aligned}$$

This implies that (x, x_{n+1}, x_{n+2}) with $x_{n+1} > 0$ cannot be a minimal solution of (P′). In other words, any minimal solution of (P′) must satisfy $x_{n+1} = 0$. By the continuity, we also see that $(x^*, x_{n+1}^*, x_{n+2}^*)$ is a minimal solution of (P′). Therefore, if a feasible solution $(\hat{x}, \hat{x}_{n+1}, \hat{x}_{n+2})$ of (P′) is minimal then $\hat{x}_{n+1} = 0$ and $c^T\hat{x} = c^T x^*$. Since **x** satisfies all the constraints of (P), it must be a minimal solution of (P). Conversely, if $(\hat{x}, 0, \hat{x}_{n+2})$ is a feasible solution of (P′), and if $\hat{x}$ is a minimal solution of (P), then the objective value $c^T\hat{x} + \kappa\hat{x}_{n+1}$ coincides with the minimal value $c^T x^* + \kappa x_{n+1}^*$. Hence it is a minimal solution of (P′). Therefore we have shown (a). We can prove (b) similarly.

If we take $x^0 = e$, $y^0 = 0$, and $z^0 = e$ then the conditions (2.30), (2.31), (2.32), and (2.33) imposed on the constants κ and λ turn out to be

$$\lambda > (e - c)^T e = n - c^T e, \tag{2.30$'$}$$

$$\kappa > 0, \tag{2.31$'$}$$

$$\lambda > (e - c)^T x^* = e^T x^* - c^T x^*, \tag{2.32$'$}$$

and

$$\kappa > (b - Ae)^T y^* = b^T y^* - y^{*T} Ae. \tag{2.33$'$}$$

§5. Concluding Remarks

Kojima [10] has proposed a method of determining optimal basic variables in Karmarkar's projective rescaling algorithm. The idea of his method is based on the relaxation technique and the duality theorem (see Sections 5 and 6 of Kojima and Tone [12]), and could be incorporated in the algorithm to determine both primal and dual optimal basic variables.

The system of equations (2.1) which has played a central role in the development of the algorithm can also be derived from the notion of a center (an analytic center) of a polytope, which has been proposed by Sonnevend [17] as a fundamental tool for a class of interior point algorithms for smooth convex minimization over polyhedra including linear programs. Since the set of the optimal solutions of (P) is bounded by the assumptions (a) and (b) of the Introduction, the polyhedron

$$S(s) = \{(\mu, x) \in R^{1+n}: x \in S, c^T x + \mu = s, \mu \geqq 0\}$$

is bounded and contains a strictly positive (μ, x) for each s greater than the optimal value s^* of (P). Hence, according to [17], we can consistently define the center of $S(s)$ to be a unique maximal solution of the problem

$$(\bar{\mathrm{P}}_s) \quad \begin{array}{l} \text{Maximize } \sum_{j=1}^{n} \log x_j + \log \mu \\ \text{subject to } (\mu, x) \in S(s), (\mu, x) > 0. \end{array}$$

Using the Karush-Kuhn-Tucker optimality condition for this problem, we can easily verify that $(\mu, x) > 0$ is a solution of $(\bar{\mathrm{P}}_s)$ for some $s > s^*$ if and only if there exists a $(y, z) \in R^{m+n}$ for which the system (2.1) is satisfied.

Similarly, the center of the polytope

$$T(t) = \{(\mu, y, z) \in R^{m+n+1} : (y, z) \in T, b^T y - \mu = t, \mu \geqq 0\},$$

which is induced from the dual problem (D), is defined to be a unique maximal solution of the problem

$$(\bar{\mathrm{D}}_t) \quad \begin{array}{l} \text{Maximize } \sum_{i=1}^{m} \log(z_i) + \log \mu \\ \text{subject to } (\mu, y, z) \in T(t), (\mu, z) > 0. \end{array}$$

Then (μ, y, z) is a maximal solution of $(\bar{\mathrm{D}}_t)$ for some $t < s^*$ if and only if there exists an $x \in R^n$ for which the system (2.1) is satisfied. Therefore the duality relations (Megiddo [14]) which we briefly explained in the Introduction could be stated in terms of the problems $(\bar{\mathrm{P}}_s)$ $(s > s^*)$ and $(\bar{\mathrm{D}}_t)$ $(t < s^*)$.

Tracing the path of centers of the polytope $T(t)$ has been employed in the polynomial time algorithm proposed by Renegar [16] and the multiplicative penalty function method by Iri and Imai [7] (see also Imai [6]).

After writing the draft of this chapter, the authors learned that Tanabe [18] has developed an interior point algorithm based on the idea of tracing the path of centers in the primal and dual spaces.

This work was presented at the conference on "Progress in Mathematical Programming" held in Monterey, California, on March 1–4, 1987. The authors learned there that, besides the papers referred in the Introduction and the comments above, many studies have been done on the paths of centers. Among others, Gonzaga [5] and Vaidya [20] both proposed algorithms that trace the path of centers to solve linear programs in $O(n^3 L)$ arithmetic operations. The algorithm given here requires $O(n^4 L)$ arithmetic operations. The algorithm will be modified and extended in [11] so as to solve a class of linear complementarity problems including linear and convex quadratic programs in $O(n^3 L)$ arithmetic operations.

Acknowledgment

The authors wish to thank Professor Masao Mori for his warm encouragement of their study on new interior point algorithms for linear programming and Dr. Nimrod Megiddo for inviting one of them to the conference on "Progress in Mathematical Programming" where this work was presented.

References

[1] I. Adler, N. Karmarkar, M. G. C. Resende, and G. Veiga, An implementation of Karmarkar's algorithm for linear programming, Working Paper, Operations Research Center, University of California, Berkeley (May 1986).

[2] K. R. Frish, The logarithmic potential method of convex programming, unpublished manuscript, University Institute of Economics, Oslo, Norway (1955).

[3] D. M. Gay, A variant of Karmarkar's linear programming algorithm for problems in standard form, *Math. Programming* **37** (1987), 81–90.

[4] P. E. Gill, W. Murray, M. A. Saunders, J. A. Tomlin, and M. H. Wright, On projected Newton barrier methods for linear programming and an equivalence to Karmarkar's projective method, *Math. Programming* **36** (1986), 183–209.

[5] C. C. Gonzaga, An algorithm for solving linear programming problems in $O(n^3 L)$ operations, in *Progress in Mathematical Programming*, N. Megiddo, (ed.), Springer-Verlag, New York, this volume.

[6] H. Imai, Extensions of the multiplicative penalty function method for linear programming, *Journal of the Operations Research Society of Japan* 30 (1987), 160–180.

[7] M. Iri and H. Imai, A multiplicative barrier function method for linear programming, *Algorithmica* **1** (1986), 455–482.

[8] N. Karmarkar, A new polynomial-time algorithm for linear programming, *Proc. 16th Annual ACM Symposium on Theory of Computing*, Washington, D.C. (1984).

[9] N. Karmarkar, A new polynomial-time algorithm for linear programming, *Combinatorica* **4** (1984), 373–395.

[10] M. Kojima, Determining basic variables of optimal solutions in Karmarkar's new LP algorithm, *Algorithmica* **1** (1986), 499–515.

[11] M. Kojima, S. Mizuno, and A. Yoshise, A polynomial-time algorithm for a class of linear complementarity problems, *Math. Programming* (to appear).

[12] M. Kojima and K. Tone, An efficient implementation of Karmarkar's new LP algorithm, Research Report No. B-180, Department of Information Sciences, Tokyo Institute of Technology, Meguro, Tokyo (April 1986).

[13] O. L. Mangasarian, *Nonlinear Programming*, McGraw-Hill, New York, 1970.

[14] N. Megiddo, Pathways to the optimal set in linear programming, this volume.

[15] N. Megiddo and M. Shub, Boundary behavior of interior point algorithms in linear programming, to appear in *Math. Oper. Res.* (1988).

[16] J. Renegar, A polynomial-time algorithm, based on Newton's method, for linear programming, *Math. Programming* **40** (1988), 59–94.

[17] G. Sonnevend, An 'analytic center' for polyhedrons and new classes of global algorithms for linear (smooth, convex) programming, *Proc. 12th IFIP Conference on System Modeling and Optimization*, Budapest (1985), to appear in *Lecture Notes in Control and Information Sciences*, Springer-Verlag, New York.

[18] K. Tanabe, Complementarity-enforcing centered Newton method for linear programming: Global method, presented at the Symposium, or "New Method

for Linear Programming," Institute of Statistical Mathematics, Tokyo (February 1987).

[19] M. J. Todd and B. P. Burrell, An extension of Karmarkar's algorithm for linear programming used dual variables, *Algorithmica* **1** (1986), 409–424.

[20] P. M. Vaidya, An algorithm for linear programming which requires $O(((m + n) \times n^2 + (m + n)^{1.5}n)L)$ arithmetic operations, AT&T Bell Laboratories, Murray Hill, N.J. (1987).

[21] H. Yamashita, A polynomially and quadratically convergent method for linear programming, Working Paper, Mathematical Systems Institute, Inc. Tokyo (October 1986).

[22] Y. Ye and M. Kojima, Recovering optimal dual solutions in Karmarkar's polynomial time algorithm for linear programming, *Math. Programming* **39** (1987), 305–317.

CHAPTER 3

An Extension of Karmarkar's Algorithm and the Trust Region Method for Quadratic Programming

Yinyu Ye

Abstract. An extension of Karmarkar's algorithm and the trust region method is developed for solving quadratic programming problems. This extension is based on the affine scaling technique, followed by optimization over a trust ellipsoidal region. It creates a sequence of interior feasible points that converge to the optimal feasible solution. The initial computational results reported here suggest the potential usefulness of this algorithm in practice.

Karmarkar's algorithm has emerged as a competitive method for solving linear programming (LP) problems (Karmarkar [11]). The algorithm is based on repeated projective transformations, followed by optimization over an inscribing sphere in the feasible region. It creates a series of interior feasible points converging to the optimal feasible solution. His work has sparked tremendous interest and has inspired others to modify his algorithm or to investigate similar methods for linear and nonlinear programming. Among those works, Barnes [1] and Vanderbei et al. [19] proposed a modification of Karmarkar's algorithm that uses the affine scaling technique by working directly in the LP standard form (see also Cavalier and Soyster [2] and Kortanek and Shi [13]), Gill et al. [7] and Megiddo [15] analyzed the barrier function method and its nice convergence pathways for linear programming (the clsssical barrier function method can be found in Frisch [6] and Fiacco and McCormick [5]), and Kapoor and Vaidya [10] and Ye [21] extended Karmarkar's projective LP algorithm to solving convex quadratic programming (QP) problems in polynomial time.

Although the affine scaling algorithm has not proved to be a polynomial-time algorithm, it works well in practice. (Since the last revision of this chapter, several papers have appeared that use variations on the barrier function method in polynomial-time algorithms; they include Gonzaga [9], Kojima et al. [12], Monteiro and Adler [16], and Ye [21].) In this chapter we first

This work was done while the author was at Integrated Systems, Inc., 2500 Mission College Blvd., Santa Clara, CA 95054, USA.

introduce the interior ellipsoid, a geometric expression of the affine scaling technique. We show how this interior ellipsoid can be used to avoid the combinatorial nature of the boundary of the polyhedron. Then, similar to the extension of Ye [21], the interior ellipsoid method for solving convex quadratic programming problems and its solution strategy and convergence behavior are discussed. We also report our initial computational results that suggest the potential usefulness of this method in practice.

§1. Convex Quadratic Programming

In this chapter we solve the following convex quadratic program:

$$\text{QP}\qquad \begin{aligned}&\text{minimize } f(x) = x^T Qx/2 + c^T x\\ &\text{subject to } x \in X = \{x\colon Ax = b,\ x \geq 0\}\end{aligned}$$

where c and $x \in R^n$, $A \in R^{m \times n}$, $b \in R^m$, $Q \in R^{n \times n}$ is symmetric and positive semidefinite, and superscript T denotes the transpose operation. The dual problem of QP can be written as

$$\text{QD}\qquad \begin{aligned}&\text{maximize } d(x, y) = b^T y - \nabla f(x)^T x + f(x) = b^T y - x^T Qx/2\\ &\text{subject to } (x, y) \in Y = \{(x, y)\colon Ax = b, A^T y \leq Qx + c, x \geq 0\}\end{aligned}$$

where $y \in R^m$ and $\nabla f(x)$ denotes the gradient vector of function $f(x)$. For all x and y that are feasible for QD,

$$d(x, y) \leq z^* \leq f(x), \tag{3.1}$$

where z^* designates the optimal objective value of QP and QD.

1.1. Optimality Conditions

Based on the Kuhn-Tucker conditions, x^* is an optimal feasible solution for QP if and only if the following three optimality conditions hold (Dantzig [4]):

(1) Primal feasibility: $x^* \in X$.
(2) Dual feasibility: $\exists y^*$ such that x^*, y^* are feasible for QD: $(x^*, y^*) \in Y$.
(3) Complementary slackness:

$$\operatorname{diag}(x^*)(Qx^* + c - A^T y^*) = 0. \tag{3.2}$$

As a result of the above conditions, if an optimal feasible solution exists for QP, then there exists a basic optimal feasible solution such that

$$\begin{pmatrix} Q & -A^T \\ A & 0 \end{pmatrix} \begin{pmatrix} x^* \\ y^* \end{pmatrix} = \begin{pmatrix} -c \\ b \end{pmatrix} \tag{3.3}$$

where $x_i^* = 0$ if $i \in I_B$, an index subset of $\{1, 2, \ldots, n\}$ (Cottle and Dantzig [3], Lemke [14], and Wolfe [20]). Generally, the nonzero components of a basic feasible solution correspond to solutions of the linear system equations with

d as the right-hand vector and B as the left-hand matrix, where d is a subvector of

$$\begin{pmatrix} -c \\ b \end{pmatrix},$$

and B is a principal submatrix of

$$\begin{pmatrix} Q & -A^T \\ A & 0 \end{pmatrix}.$$

1.2. Assumptions

In this chapter we assume that there exists an interior feasible solution x^0 for QP with

A1 $\quad x^0 > 0$

and we further make an implicit assumption that the optimal solution can be found is a bounded polyhedron, that is,

A2 $\quad x \leq Me,$

where e is the vector of all ones, and $0 < M < \infty$. Assumptions A1 and A2 essentially say that the feasible polyhedron of X is bounded and closed and that it contains a nonempty interior region.

§2. Interior Ellipsoid: A Geometric Expression of the Affine Scaling Technique

To illustrate the basic concept of the proposed interior ellipsoid method, we use a linear objective function. Figure 3.1 displays a feasible polytope with the arrow pointing in the descent direction of the objective function. In the pivot

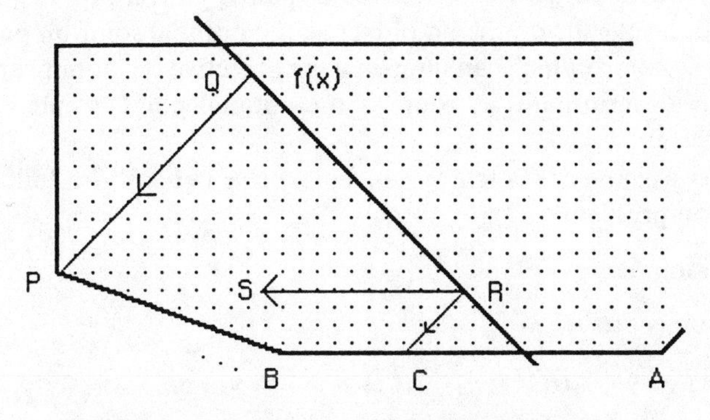

Figure 3.1. Objective contour and feasible polytope.

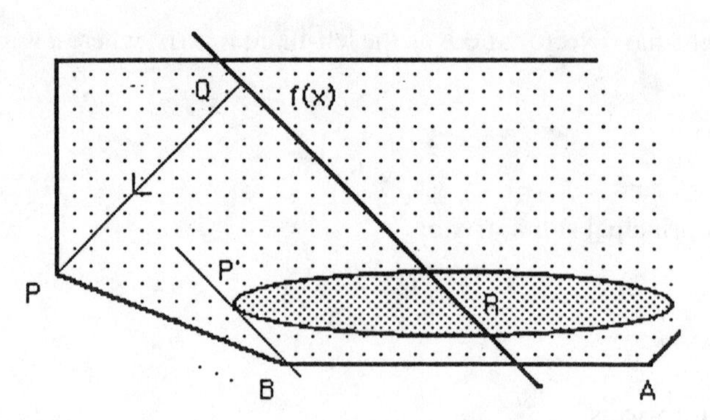

Figure 3.2. The interior ellipsoid approach.

method [3, 14, 20], the solution moves from vertex to vertex, that is, from A to $B, \ldots,$ converging to the optimal solution point P in a finite number of steps.

In the gradient-projection method of Rosen [17], the solution may start from the interior of the polytope. As illustrated, if the starting point happened to be at point Q, then the method could reach point P in one step. However, if we started from point R, the method would generate the boundary point C and then would move along the edges of the polytope. As soon as the iterative solution reached the boundary, a combinatorial decision would have to be made to re-form the active constraint set at each step. In the worst case, the optimal solution would be reached in an exponential number of steps.

The issue is: how can we avoid hitting the "wrong" boundary? In other words, can we develop a mechanism to move the solution in the direction R–S instead of R–C?

A geometric expression derived from the concept of the trust region method (see Goldfeld et al. [8] and Sorensen [18]) can be drawn as an interior ellipsoid centered at the starting point R in the feasible polytope, shown in Figure 3.2. Then the objective function can be minimized over this interior ellipsoid to generate the next interior solution point P'. A series of such ellipsoids can thus be constructed to generate a sequence of points converging to the optimal solution point that sits on the boundary. If the optimal solution point itself is an interior solution (which can happen if the objective is a nonlinear function), then the series terminates as soon as the optimal point is encircled by the newest ellipsoid.

The above geometric picture can be represented by the following sub-optimization problem:

$$\text{QPS} \qquad \begin{aligned} &\text{minimize } f(x) \\ &\text{subject to } Ax = b \\ &\qquad \|D^{-1}(x - x^k)\|^2 \le \beta^2 < 1 \end{aligned}$$

where D is an invertible diagonal matrix and x^k is the interior feasible solution given at the beginning of the kth iteration. The last constraint, $\|D^{-1}(x - x^k)\|^2 \leq \beta^2$, corresponds to an ellipsoid embedded in the positive orthant $\{x \in R^n : x \geq 0\}$. Therefore $\{x\colon Ax = b,\ \|D^{-1}(x - x^k)\|^2 \leq \beta^2\}$ is an algebraic representation of the interior ellipsoid centered at x^k in X of QP. The parameter $\beta > 0$ characterizes the size of the ellipsoid, and D affects the orientation and shape of the ellipsoid. In the affine scaling algorithm,

$$D = \operatorname{diag}(x^k). \tag{3.4}$$

With this D, the ellipsoid constraint can be rewritten as

$$\|D^{-1}x - e\| \leq \beta < 1. \tag{3.5}$$

Inequality (3.5) implies $x > 0$; that is, any feasible solution for QPS is a positive (interior) feasible solution for QP. In other words, the nonnegativity constraint $x \geq 0$ is redundant in QPS. Thus, the algorithm can be described as

Algorithm

At the kth iteration do
  begin
    $D = \operatorname{diag}(x^k)$;
    let x^{k+1} be the minimal solution for QPS;
    $k = k + 1$;
  end.

§3. Convergence Analysis

In order to analyze the convergence of the algorithm, we derive the optimality conditions for QPS. First, we note that the first-order conditions completely characterize the optimality of QPS since the objective is a convex function and the feasible region is a convex set. Let x^{k+1} be the optimal solution, y^{k+1} be the optimal Lagrange multiplier for the equality constraints, and μ^{k+1} be the optimal Lagrange multiplier for the ellipsoid constraint of QPS. Then x^{k+1}, y^{k+1}, and μ^{k+1} meet the following equalities and inequalities:

$$\nabla f(x^{k+1}) - A^T y^{k+1} + \mu^{k+1} D^{-2}(x^{k+1} - x^k) = 0, \tag{3.6}$$

$$Ax^{k+1} = b, \tag{3.7}$$

$$\|D^{-1}(x^{k+1} - x^k)\|^2 \leq \beta^2 \quad \text{and} \quad \mu^{k+1} \geq 0, \tag{3.8}$$

and

$$\mu^{k+1}(\beta^2 - \|D^{-1}(x^{k+1} - x^k)\|^2) = 0. \tag{3.9}$$

Analytically, from (3.6) and (3.7) we have

$$(AD^2A^T)y^{k+1} = AD^2\nabla f(x^{k+1}). \tag{3.10}$$

Let

$$p^k = D(\nabla f(x^{k+1}) - A^T y^{k+1}); \tag{3.11}$$

then from (3.6)

$$\mu^{k+1} = \frac{\|p^k\|}{\|D^{-1}(x^{k+1} - x^k)\|}, \tag{3.12}$$

$$\nabla f(x^{k+1})^T D p^k = \|p^k\|^2, \tag{3.13}$$

and, if $p^k \neq 0$, from (3.6) and (3.9)

$$\|D^{-1}(x^{k+1} - x^k)\| = \beta \tag{3.14}$$

and

$$x^{k+1} = x^k - \beta \frac{Dp^k}{\|p^k\|}. \tag{3.15}$$

Here p^k represents the complementary slackness vector of (3.2) at the kth iteration.

Now we prove the following three lemmas, which are similar to the analysis of the affine scaling algorithm as described in several papers (Barnes [1] and Vanderbei et al. [19]). However, because of the nonlinearity of the objective function, some results and different from those presented so far. The first lemma basically states that if the algorithm "jams," then it arrives at a positive (interior) optimal feasible solution in a finite number of iterations. (This case would never happen if the objective was a linear function.)

Lemma 3.1. *If $p^k = 0$ (i.e., $\mu^{k+1} = 0$) for $k < \infty$, then x^{k+1} and y^{k+1} are optimal for* QP *and* QD.

PROOF. Note from (3.11) that

$$p^k = 0$$

implies

$$D(\nabla f(x^{k+1}) - A^T y^{k+1}) = 0,$$

which implies

$$(\nabla f(x^{k+1}) - A^T y^{k+1}) = 0 \tag{3.16}$$

and

$$\operatorname{diag}(x^{k+1})(\nabla f(x^{k+1}) - A^T y^{k+1}) = 0. \tag{3.17}$$

Therefore, the conclusion in Lemma 3.1 follows from the optimality conditions:

(1) x^{k+1} is feasible for QP,
(2) x^{k+1} and y^{k+1} are feasible for QD from (3.16), and
(3) complementary slackness is satisfied from (3.17).

If $\|p^k\| > 0$ for all finite k, the second lemma claims that $\|p^k\| \to 0$, where $\to$ designates "converges to." In addition, if $f(x)$ is strictly convex, then x^k converges. This property is relatively new to the convergence of the affine scaling algorithm for linear programming.

Lemma 3.2. *Let $\|p^k\| > 0$ for all $k < \infty$ and the optimal objective value of* QP *be bounded from below. Then $\|p^k\| \to 0$; furthermore, if the Hessian Q is positive definite, then x^k converges.*

Proof. Using (3.13) and (3.15), we have

$$\begin{aligned} f(x^k) &= f\left(x^{k+1} + \beta \frac{Dp^k}{\|p^k\|}\right) \\ &= f(x^{k+1}) + \frac{\beta}{\|p^k\|} \nabla f(x^{k+1}) Dp^k + \frac{\beta^2}{2\|p^k\|^2}(p^k)^T DQDp^k \\ &= f(x^{k+1}) + \beta\|p^k\| + \frac{\beta^2}{2\|p^k\|^2}(p^k)^T DQDp^k, \end{aligned} \tag{3.18}$$

where the Hessian Q is at least positive semidefinite. Therefore,

$$\beta\|p^k\| \le f(x^k) - f(x^{k+1}).$$

Since $f(x^k)$ is monotonically decreasing and is bounded from below, $f(x^k)$ must converge and $f(x^k) - f(x^{k+1}) \to 0$, which implies $\|p^k\| \to 0$. Moreover, via (3.18),

$$\frac{\beta^2}{2\|p^k\|^2}(p^k)^T DQDp^k \le f(x^k) - f(x^{k+1}).$$

Hence, if Q is positive definite, it follows that

$$\frac{\|Dp^k\|}{\|p^k\|} \to 0,$$

which implies from (3.15) that

$$\|x^k - x^{k+1}\| \to 0. \tag{3.19}$$

As the feasible polytope is bounded and closed, $\{x^k\}$ must have a subsequence converging to x^∞. This fact and (3.19) imply that the whole sequence $\{x^k\}$ converges to x^∞.

The third lemma establishes that if x^k and y^k converge, then they converge to the feasible solution for QD. This fact is proved for LP by several authors mentioned above, and it is also true if the objective function is nonlinear.

Lemma 3.3. *If $\|p^k\| > 0$ for all $k < \infty$, $x^k \to x^\infty$, $y^k \to y^\infty$, and $\|p^k\| \to 0$, then x^∞ and y^∞ are feasible for* QD.

PROOF. We have

$$p_i^\infty = x_i^\infty(\nabla f(x^\infty) - A^T y^\infty)_i = 0 \quad \text{for } i = 1, 2, \ldots, n. \tag{3.20}$$

Suppose y^∞ is not feasible for QD, that is, $\exists \varepsilon > 0$ and $1 \leq j \leq n$ such that

$$(\nabla f(x^\infty) - A^T y^\infty)_j \leq -\varepsilon < 0;$$

then $\exists K > 0$ such that for all $\infty > k > K$

$$(\nabla f(x^{k+1}) - A^T y^{k+1})_j < -\frac{\varepsilon}{2}.$$

At the kth ($k > K$) iteration of the algorithm,

$$x_j^{k+1} = x_j^k \left(1 - \beta \frac{x_j^k(\nabla f(x^{k+1}) - A^T y^{k+1})_j}{\|p^k\|}\right) > x_j^k,$$

hence

$$x_j^{k+1} > x_j^k > x_j^K > 0 \quad \text{for all } k > K.$$

Thus, $\{x_j^k\}$ is a strictly increasing positive series for $k > K$. Since neither x_j^k nor $(\nabla f(x^k) - A^T y^k)_j$ converges to 0, $x_j^k(\nabla f(x^k) - A^T y^k)_j$ does *not* converge to 0. This contradicts (3.20). Therefore, it must be true that

$$\nabla f(x^\infty) - A^T y^\infty \geq 0,$$

that is, x^∞ and y^∞ are feasible for QD.

Note that y^k always exists as a solution of (3.10), even though it may not be unique. Therefore, even if x^k converges, y^k does not necessarily converge. However, if x^k converges to a nondegenerate solution (with not less than m nonzero components), then y^k converges as the unique solution of (3.10). This observation and the three lemmas lead to the following convergence theorem.

Theorem 3.1. *Let the optimal objective value of* QP *be bounded from below and the optimal solution be nondegenerate, and let the Hessian Q be positive definite. Then the algorithm generates solution sequences that converge to the optimal solutions for both* QP *and* QD.

PROOF. If $p^k = 0$ for $k < \infty$, the conclusion of Theorem 3.1 follows from Lemma 3.1; otherwise, from Lemma 3.2, Lemma 3.3, and (3.10), x^k and y^k converge and x^∞ and y^∞ are feasible for QP and QD, and from Lemma 3.2, $\|p^\infty\| = 0$, which implies that complementary slackness is satisfied. Thus, x^∞ and y^∞ are optimal for QP and QD.

Theorem 3.1 applies to strictly convex quadratic programming. In general, we have the following theorem based on the three lemmas.

Theorem 3.2. *Let the optimal objective value of* QP *be bounded from below and let x^k and y^k converge. Then the algorithm generates solution sequences x^k and y^k that converge to the optimal solutions for both* QP *and* QD.

Theorem 3.2 agrees to the result obtained for the affine algorithm of linear programming. To evaluate the local convergence rate of the algorithm, we have

Lemma 3.4. *Let x^{k+1} and y^{k+1} be feasible for* QD. *Then*

$$f(x^{k+1}) - z^* \le \left(1 - \frac{\beta}{\sqrt{n}}\right)(f(x^k) - z^*).$$

PROOF. Since $f(x)$ is a convex function,

$$f(x^k) - f(x^{k+1}) \ge \nabla f(x^{k+1})^T(x^k - x^{k+1}). \tag{3.21}$$

We are also given that y^{k+1} is feasible for QD, so from (3.1)

$$d(x^{k+1}, y^{k+1}) - z^* = f(x^{k+1}) - \nabla f(x^{k+1})^T x^{k+1} + b^T y^{k+1} - z^* \le 0. \tag{3.22}$$

From (3.21) and (3.22)

$$\nabla f(x^{k+1})^T x^k - b^T y^{k+1} + d(x^{k+1}, y^{k+1}) - z^* \le f(x^k) - z^*. \tag{3.23}$$

According to (3.13) and (3.15),

$$\nabla f(x^{k+1})^T x^{k+1} = \nabla f(x^{k+1})^T x^k - \beta \| p^k \|. \tag{3.24}$$

Using Hölder's inequality and (3.11), and noting that $p^k \ge 0$,

$$\begin{aligned} \| p^k \| &\ge \frac{1}{\sqrt{n}}\left(\sum_{i=1}^{n} |p_i^k|\right) \\ &= \frac{1}{\sqrt{n}}\left(\sum_{i=1}^{n} (p_i^k)\right) \\ &= \frac{1}{\sqrt{n}}(\nabla f(x^{k+1})^T x^k - b^T y^{k+1}). \end{aligned} \tag{3.25}$$

Because of (3.22)–(3.25)

$$\begin{aligned} f(x^{k+1}) - z^* &= \nabla f(x^{k+1})^T x^{k+1} - b^T y^{k+1} + d(x^{k+1}, y^{k+1}) - z^* \\ &\le \nabla f(x^{k+1})^T x^k - \beta \| p^k \| - b^T y^{k+1} \\ &\quad + \left(1 - \frac{\beta}{\sqrt{n}}\right)(d(x^{k+1}, y^{k+1}) - z^*) \\ &\le \left(1 - \frac{\beta}{\sqrt{n}}\right)(\nabla f(x^{k+1})^T x^k - b^T y^{k+1} + d(x^{k+1}, y^{k+1}) - z^*) \\ &\le \left(1 - \frac{\beta}{\sqrt{n}}\right)(f(x^k) - z^*). \end{aligned}$$

According to Theorem 3.1, y^k does converge to a dual feasible solution. Therefore, Lemma 3.4 implies the following corollary.

Corollary 3.1. *Let x^k and y^k converge in the algorithm. Then the local (asymptotic) convergence ratio of the algorithm is $(1 - O(1/\sqrt{n}))$.*

Lemma 3.4 and Corollary 3.1 are of limited value since dual feasibility is typically way off until the algorithm is homing in on an optimal basic solution. In Section 5 we will discuss how to obtain a dual feasible solution sooner in the course of the algorithm. Now the question is how to solve QPS of the suboptimization problem.

§4. How to Solve QPS

The question is equivalent to finding x^{k+1}, y^{k+1}, and μ^{k+1} that satisfy optimality conditions (3.6)–(3.9). Actually, optimality conditions (3.6) and (3.7) can be written in a matrix form

$$\begin{pmatrix} Q + \mu D^{-2} & -A^T \\ A & 0 \end{pmatrix} \begin{pmatrix} x \\ y \end{pmatrix} = \begin{pmatrix} -c + \mu D^{-1} e \\ b \end{pmatrix}. \tag{3.26}$$

The system of linear equations (3.26) can be solved by approximating the multiplier μ. To do this one-dimensional search, we first develop a bound for μ^{k+1} in the following lemma.

Lemma 3.5.

$$0 \le \mu^{k+1} \le \frac{f(x^k) - f(x^{k+1})}{\beta^2} \quad \text{for all } k < \infty.$$

Moreover, if x^{k+1} and y^{k+1} are feasible for QP and QD,

$$\frac{f(x^{k+1}) - z^*}{\beta\sqrt{n}} \le \mu^{k+1} \le \frac{f(x^k) - f(x^{k+1})}{\beta^2}.$$

Proof. From (3.21) and (3.12)–(3.15),

$$\begin{aligned} f(x^k) - f(x^{k+1}) &\ge \nabla f(x^{k+1})^T (x^k - x^{k+1}) \\ &= \beta \nabla f(x^{k+1})^T D \frac{p^k}{\|p^k\|} \\ &= \beta \|p^k\| \\ &= \beta^2 \mu^{k+1}. \end{aligned}$$

Additionally, according to (3.22)–(3.25),

$$\begin{aligned} \beta\mu^{k+1} &= \|p^k\| \\ &\ge \frac{1}{\sqrt{n}} (\nabla f(x^{k+1})^T x^k - b^T y^{k+1}) \end{aligned}$$

$$\geq \frac{1}{\sqrt{n}}(\nabla f(x^{k+1})^T x^{k+1} - b^T y^{k+1})$$

$$\geq \frac{f(x^{k+1}) - z^*}{\sqrt{n}}$$

if x^{k+1} and y^{k+1} are feasible for QP and QD.

Lemma 3.5 indicates that μ^k is well bounded and that it converges to zero. For any given μ, we can solve (3.26) to obtain $x(\mu)$ and $y(\mu)$. Then we check to see if $x(\mu) > 0$. If $x(\mu) > 0$, then we return the μ, $x(\mu)$, and $y(\mu)$ as μ^{k+1}, x^{k+1}, and y^{k+1}. Here we relax the restriction of $\beta < 1$ to achieve practical efficiency. This enlargement of the size of the ellipsoid is valid as long as x^{k+1} is a positive (interior) feasible solution for QP. Obviously, from Lemma 3.5, $x(O(f(x^k) - z^*)) > 0$. On the other hand, if $x(0) > 0$, then we obtain a positive (interior) optimal feasible solution for QP from Lemma 3.1. Overall, the above process can be implemented by using the binary line-search technique, as shown in the following procedure.

Procedure 3.1

Begin

1. Set $\mu^{k+1} = O(f(x^{k-1}) - f(x^k))$
2. Solve (3.26) to obtain x and y
3. Check to see if $x > 0$. If "yes," then stop and return $x^{k+1} = x$, $y^{k+1} = y$, and μ^{k+1}
   else $\mu^{k+1} = 2\mu^{k+1}$ and go to 2

end.

Essentially, μ characterizes the size (β) of the ellipsoid from (3.12). Consequently, as in the trust region method, searching for $\mu^{k+1} \in \{\mu\colon x(\mu) > 0\}$ is equivalent to searching for a valid size of the interior trust ellipsoidal region. Procedure 3.1 may be terminated in several steps, and each step solves a system of linear equations (3.26). However, since μ^{k+1} is well bounded by Lemma 3.5 and the size β of the ellipsoid is quite flexible, Procedure 3.1 is usually terminated in one or two steps in our computational experience. Moreover, since there is no need to solve QPS exactly, one can use some existing iterative algorithms such as the conjugate direction method to achieve further practical efficiency.

§5. Computational Experiment and Further Discussion

In this section we report our initial computational experiment made to elucidate the theoretical analysis in Sections 3 and 4. Before we go into the numerical results, one may ask, what is the stopping criterion to terminate the algorithm?

5.1. Dual Feasibility and Stopping Rule

One choice is using the condition $\|p^k\| \leq \varepsilon$. As noticed by many authors (for example, Vanderbei et al. [19]), $\|p^k\| = 0$ at every vertex of the feasible region; hence it shouldn't be used as the sole criterion unless y^{k+1} becomes feasible for QD. This raises the same question as the one posed at the end of Section 3: how can one obtain a dual feasible solution in the course of the algorithm?

From the barrier function analyzed by Gill et al. [7] and Megiddo [15], we actually solve the following system of linear equations, instead of (3.26), in Procedure 3.1:

$$\begin{pmatrix} Q + \mu D^{-2} & -A^T \\ A & 0 \end{pmatrix} \begin{pmatrix} x \\ y \end{pmatrix} = \begin{pmatrix} -c + (1 + \alpha/\sqrt{n})\mu D^{-1} e \\ b \end{pmatrix}. \tag{3.27}$$

The solutions x^{k+1}, y^{k+1}, and μ^{k+1} obtained by solving (3.27) in Procedure 3.1 essentially represent the optimal solutions for the problem

$$\begin{aligned} &\text{minimize } f(x) - (\mu^{k+1}\alpha/\sqrt{n})e^T D^{-1}(x - x^k) \\ \text{QPS} \quad &\text{subject to } Ax = b \\ &\|D^{-1}(x - x^k)\|^2 \leq \beta^2 < 1 \end{aligned}$$

where $\alpha < 1$ and $e^T D^{-1}(x - x^k)$ is the first-order approximation of the barrier function $\sum(\ln x_i)$ at x^k. With this dynamical choice of the barrier parameter $\mu^{k+1}\alpha/\sqrt{n}$, one can verify that the main convergence results of the algorithm are still valid. However, the barrier term helps to avoid hitting the nonoptimal vertex and to generate the dual feasible solution. In fact, the termination criterion used in our computational experiment is

$$(x^k, y^k) \in Y \quad \text{and} \quad f(x^k) - d(x^k, y^k) \leq 10^{-4}\max(1, |f(x^k)|).$$

This criterion ensures that the obtained primal and dual *feasible* objective values have the accuracy over five digits of the optimal one.

5.2. Test QP Problems

We carried out the computational experiment as follows: random QP problems of various dimensions were generated by choosing each entry of A, c, and an n-vector x as a uniform [0 1] random variable independently. Another matrix, $P_{n \times m}$, was also randomly generated, and then the assignments $Q = PP^T$ and $b = Ax$ were made. We used the Phase 1 technique of LP (Ye and Kojima [22]) to find the initial interior feasible solution x^0, then used the algorithm with Procedure 3.1 (solving system (3.27)), called Phase 2, to compute the QP problem. The program was written in MATRIX$_x$™ of Integrated Systems Inc.

For each size of problem $m \times n$ we generated several random problems and recorded the number of iterations (Phase 2), ITNS, and the number of

Table 3.1. Computational Results of the IE Method for Solving Random Convex QP Problems

Problem dimension	ITNS	LSYS
5 × 10	11	12
10 × 20	12	13
15 × 30	12	14
20 × 40	12	14
25 × 50	13	14
30 × 60	13	15
35 × 70	14	15
40 × 80	14	15
50 × 100	14	16

systems of linear equations (3.27) solved, LSYS. The worst results are given in Table 3.1. We also used this program to solve small *real* problems, resulting in behavior similar to that in Table 3.1.

Overall, the computational results can be summarized as follows:

(1) The number of iterations of this approach is insensitive to the size of QP problems.
(2) The number of linear systems solved is almost identical to the number of iterations. This fact shows that each iteration solves about one system of linear equations.
(3) The algorithm usually starts to generate the dual feasible solution after few iterations.

5.3. Indefinite Case

We also tried this program to solve randomly generated indefinite and negative definite QP problems. The program seems to generate a local minimum (both primal and dual) in about the same number of iterations as for the positive semidefinite case. As we discussed above, each iteration of the algorithm resembles the trust region method. Therefore, the decent property of the trust region method is preserved in our approach for general QP problems. However, no solid theoretical result is analyzed for indefinite or negative definite cases.

5.4. Sparsity

The last computational issue is to exploit the sparsity in solving (3.26) or (3.27). As we noticed, the sparsity structure of the linear system (3.26) remains unchanged during the entire iterative process. Therefore, the solution time can be significantly shortened by using sparsity code for solving the system of linear equations.

References

[1] E. R. Barnes, A variation on Karmarkar's algorithm for solving linear programming, *Math. Programming* **36** (1986), 174–182.

[2] T. M. Cavalier and A. L. Soyster, Some computation experience and a modification of the Karmarkar algorithm, ISME Working Paper 85-105, Department of Industrial and Management Systems Engineering, Pennsylvania State University (1985).

[3] R. W. Cottle and G. B. Dantzig, Complementary pivot theory of mathematical programming, *Linear Algebra Appl.* **1** (1968), 103–125.

[4] G. B. Dantzig, *Linear Programming and Extensions*, Princeton University Press, Princeton, N.J., 1963.

[5] A. V. Fiacco and G. P. McCormick, *Nonlinear Programming: Sequential Unconstrained Minimization Techniques*, Wiley, New York, 1968.

[6] K. R. Frisch, The logarithmic potential method of convex programming, Memorandum, University Institute of Economics, Oslo, Norway (1955).

[7] P. E. Gill, W. Murray, M. A. Saunders, J. A. Tomlin, and M. H. Wright, On projected Newton barrier methods for linear programming and an equivalence to Karmarkar's projective method, *Math. Programming* **36** (1986), 183–209.

[8] S. M. Goldfeld, R. E. Quandt, and H. F. Trotter, Maximization by quadratic hill climbing, *Econometrica* **34** (1966), 541–551.

[9] C. C. Gonzaga, An algorithm for solving linear programming problems in $O(n^3L)$ operations, Memorandum No. UCB/ERL M87/10, Electronic Research Laboratory, University of California, Berkeley (1987).

[10] S. Kapoor and P. Vaidya, Fast algorithms for convex quadratic programming and multicommodity flows, *Proc. 18th Annual ACM Symposium on Theory of Computing* (1986), 147—159.

[11] N. Karmarkar, A new polynomial-time algorithm for linear programming, *Combinatorica* **4** (1984), 373–395.

[12] M. Kojima, S. Mizuno, and A. Yoshise, A polynomial-time algorithm for a class of linear complementarity problems, Research Report, Department of Information Sciences, Tokyo Institute of Technology (1987).

[13] K. O. Kortanek and M. Shi, Convergence results and numerical experiments on a linear programming hybrid algorithm, *European Journal of Operational Research* (to appear).

[14] C. E. Lemke, Bimatrix equilibrium points and mathematical programming, *Management Sci.* **11** (1965), 681–689.

[15] N. Megiddo, Pathways to the optimal set in linear programming. In: *Progress in Mathematical Programming: Interior-Point and Related Methods*. Springer-Verlag, New York (1989).

[16] R. C. Monteiro and I. Adler, An $O(n^3L)$ interior point algorithm for convex quadratic programming, Manuscript, Department of Industrial Engineering and Operations Research, University of California, Berkeley (1986).

[17] J. Rosen, The gradient projection method for nonlinear programming, I. Linear constraints, *J. Soc. Ind. Appl. Math.* **12** (1964), 74–92.

[18] D. C. Sorensen, Trust region methods for unconstrained minimization, in *Nonlinear Optimization*, M.J.D. Powell (ed.), Academic Press, London, 1981.

[19] R. J. Vanderbei, M. S. Meketon, and B. A. Freedman, On a modification of Karmarkar's linear programming algorithm, *Algorithmica* **1** (1986), 395–407.

[20] P. Wolfe, The simplex algorithm for quadratic programming, *Econometrica* **27** (1959), 382–398.

[21] Y. Ye, Interior algorithms for linear, quadratic, and linearly constrained convex programming, Ph.D. Thesis, Department of Engineering-Economic Systems, Stanford University (1987).

[22] Y. Ye and M. Kojima, Recovering optimal dual solutions in Karmarkar's algorithm for linear programming, *Math. Programming* **39** (1987), 305–317.

CHAPTER 4

Approximate Projections in a Projective Method for the Linear Feasibility Problem

Jean-Philippe Vial

Abstract. The key issue in implementing a projective method is the projection operation. In order to cut down computations, several authors have suggested using approximations instead of the projection itself. However, these approximations are not directly compatible with the standard proofs of polynomial complexity. In this chapter we present a relaxation of the main convergence lemma which makes it possible to accommodate approximate projections. We propose several types of approximations that preserve polynomial complexity for the version of Karmarkar's algorithm presented by de Ghellinck and Vial.

§1. Introduction

The new projective algorithm of Karmarkar for linear programming [12] has generated considerable interest. It is polynomial in complexity and experiments show that the number of iterations that are required to yield a solution is remarkably low. However, this feature is not sufficient in itself to guarantee that projective algorithms will outperform the simplex method because each iteration involves a costly operation, namely a projection on the null space of a matrix. This adverse factor may ruin the overall efficiency of the algorithm unless highly efficient implementations are used.

In order to cut down computations several authors [1, 5–8, 13, 14, 16–18] have proposed using approximations of the projection instead of the projection itself. The idea is certainly appealing, and very encouraging numerical results have been reported in the literature. However, standard convergence proofs of polynomial convergence rely on exact projections. It is relevant to see whether these proofs can accommodate approximate projections. More-

This work has been completed under partial support of research grant 1.467.0.86 of the Fonds National Suisse de la Recherche Scientifique.

over, studying approximations in the framework of complexity gives better insight into the type of approximation that is really sought and its required level of accuracy.

In this chapter, approximations and their impact on the complexity of projective algorithms are analyzed in the framework of the method of de Ghellinck and Vial [2]. That method deals with the problem of solving a set of linear equations on the nonnegative orthant, a problem which we shall henceforth call the "feasibility problem." The algorithm is a constrained Newton method to minimize a potential that is the difference of two terms: the logarithm of the violation of the constraints and the sum of the logarithms of the individual coordinates of the current iterate. The constrained Newton direction is obtained by a projection operation. Linear programs with known optimal objective function value can be solved by the method [2] since they are equivalent to feasibility problems with a special constraint that equates the objective function value to the optimal value.

The convergence analysis in [2] (see also [3]) is based on a convergence lemma which shows that a fixed minimal decrease of the potential can be achieved by a search along the exact projection. The equivalence with Karmarkar's algorithm is proved in [2].

The main result of this chapter is that the convergence lemma can be relaxed. It makes it possible to use approximate projections instead of exact projections, provided these approximations meet some accuracy tolerances that are specified in the lemma. In particular, it appears that the search direction need not be exactly in the null space of the scaled constraints of the problem. The possible degradation it introduces can be measured and controled by the first term of the potential.

Our paper is closely related to [7], which also deals with linear programs with known optimal value. Approximations in the null space that preserve polynomiality are proposed and practical ways of computing them are discussed. The setting in [7] is the same as the original algorithm of Karmarkar. The paper [8] extends [7] to linear programs with unknown optimal value.

In the growing body of literature dealing with projective algorithms, many proposals for approximations have been made. Concerning partial updating, Karmarkar himself [12] has proposed a modified algorithm, for which he establishes improved polynomial complexity. Numerical experiments with this approach and some variants are reported in [1, 5, 17]. Iterative methods are discussed and/or implemented in [1, 6–8, 13, 14]. The complexity issue raised by approximations is discussed in [7, 8, 12].

For computing the projections, especially for a large sparse system, we refer to [9, 11, 15] and to the discussion of [20]. Finally, we point out that the use of iterative methods for solving the least squares problem is very reminiscent of what is done in large sparse nonlinear programming [4, 19, 21].

The paper is organized as follows. In Section 2 the solution method is discussed and the basic algorithm is presented. In Section 3 a relaxed version of the main convergence lemma is stated and proved. In Section 4 two types of approximations of the projection are presented, each solving one subset of the least squares equations and approximating the other. Such approximations can be obtained by an iterative method such as the conjugate gradient method. Stopping criteria for the iterative method are given. In Section 5 another approach is discussed which mixes partial updating and approximations.

In this chapter we use the following notation. If x is a vector in $\Re^n$, x' is its transpose, x_j its jth component, and X the $n \times n$ diagonal matrix with diagonal entries x_j. The vector $\mathbf{1} = (1, \ldots, 1)'$ has all its components equal to 1. Its dimension, unless specified, has to be inferred from the context. The Euclidean norm is $\|x\| = [\sum_{j=1}^{n} (x_j)^2]^{1/2}$ and the l_∞ norm $\|x\|_\infty = \max_j |x_j|$. If A is an $m \times n$ matrix, A' is its transpose.

§2. A Projective Method for the Linear Feasibility Problem

We consider the following linear feasibility problem on the rationals

$$\text{(F)} \qquad Bx = b, \qquad x \in \Re^n_+,$$

where B is an $n \times n$ matrix and $b \in \Re^m$. $\Re^n_+$ is the nonnegative orthant in $\Re^n$. We associate with F the homogeneous problem

$$(\tilde{\text{F}}) \qquad A\tilde{x} = 0, \qquad \tilde{x} \geq 0, \tilde{x} \in \Re^{n+1},$$

where $A = (-b, B)$. Note that the indices of $\tilde{x} \in \Re^{n+1}$ run from 0, 1, ... to n.

If x solves F, $\tilde{x} = (1, x')'$ is a nontrivial solution to $\tilde{F}$. Conversely, any nontrivial solution of $\tilde{F}$ such that $\tilde{x}_0 \neq 0$ can be scaled to $\tilde{x} = (1, x')'$, where x solves F. We introduce the assumption (see [3, 10]):

Assumption A. The system

$$(\text{F}_0) \qquad Bx = 0, \qquad x \geq 0, x \in \Re^n$$

has no solution other than the trivial solution $x = 0$.

Assumption A essentially amounts to the boundedness of the solution set in F, a nonrestrictive assumption. Under this assumption it is clear that any nontrivial solution to $\tilde{F}$ is such that $x_0 \neq 0$. We shall assume Assumption A throughout the chapter and we shall concentrate on the problem of finding a nontrivial solution to $\tilde{F}$. For the sake of simplicity we shall drop the notation $\tilde{x}$. In the sequel the vector x will be an element in $\Re^{n+1}$. Whenever x, with $x_0 \neq 0$, solves $\tilde{F}$ it will be implicit that a solution to F is readily obtained by a simple scaling.

We introduce the potential

$$\phi(x) = (n+1)\ln\|Ax\| - \sum_{i=0}^{n} \ln x_i, \tag{4.1}$$

which is defined on the interior of $\Re_+^{n+1}$. This function has the property that for $\varepsilon > 0$ small enough (where small can be made precise in terms of the input size of the problem), the condition $\exp\{\phi(x)\} < \varepsilon$ implies that an exact solution to F can be computed from x in polynomial time (see [3]). The algorithm in [2] is aimed at minimizing ϕ.

The nonconvex problem of minimizing ϕ over int $\Re_+^{n+1}$ can be replaced by a simpler convex problem. Let $x \in \text{int } \Re_+^{n+1}$ be some fixed point. We associate to it the problem in the variable $p \in \Re^{n+1}$

$$\underset{p}{\text{Minimize}}\ \{\phi(x+p) \mid Ap = 0,\ x_j + p_j > 0,\ j = 0, 1, \dots n\}.$$

Since $A(x + p) = Ax$ this problem is equivalent to

$$\text{(NLP)} \qquad \underset{p}{\text{Maximize}} \left\{ \sum_{i=0}^{n} \ln(x_j + p_j) \mid Ap = 0,\ x_j + p_j > 0,\ j = 0, 1, \dots n \right\}.$$

The objective is clearly concave in the variable p.

The projective algorithm is a constrained Newton-like method for solving NLP. The constrained Newton direction is the solution to

$$\text{(QP)} \qquad \underset{p}{\text{Minimize}}\ \{\|X^{-1}p - \mathbf{1}\|^2 \mid Ap = 0\}.$$

Introducing the change of variable $p = Xq$, one easily shows that the optimal solution of QP can be obtained by solving the least squares equations in the unknowns $q \in \Re^{n+1}$ and $u \in \Re^m$

$$q + XA'u = \mathbf{1}, \tag{4.2}$$

$$AXq = 0. \tag{4.3}$$

The solution in q is the projection of the vector $\mathbf{1}$ on the null space of the matrix AX.

Finally, a step of size α is taken along the direction p so as to solve approximately the one-dimensional minimization problem

$$\text{(LS)} \qquad \underset{\alpha}{\text{Minimize}}\ \{\phi(x + \alpha p) \mid x_j + \alpha p_j > 0,\ j = 0, 1, \dots n\}.$$

This is the essence of the algorithm that we state below.

Algorithm 4.1

Initialization. Let $\varepsilon > 0$ be a tolerance level. Let $x^0 = \mathbf{1} \in \Re^{n+1}$ and let $k = 0$.

Basic iteration. $x^k \in \text{int } \Re_+^{n+1}$ is given and satisfies with $x_0^k = 1$.

1. *Projection step.* Find the projection $q \in \Re^{n+1}$ of $\mathbf{1}$ onto the null space of AX_k. This is done by solving the least squares equations (4.2) and (4.3) with X_k instead of X.
2. *Convergence tests.*
   2.1. (*infeasibility test*) If $q_{\max} = \max_j q_j < 1$, STOP: F has no feasible solution.

2.2. (*feasibility test*) If $q_{\min} = \min_j q_j \geq 0$,
STOP: The point x, with $x_j = (q_0)^{-1}(X_k q)_j$, $j = 0, 1, \ldots, n$, solves F.

2.3. (ε-*convergence*) If $\exp\{\phi(x^k)\} < \varepsilon$
STOP: x^k is an ε-solution to F.

3. *Line search.* Find $\alpha \in \mathfrak{R}_+$ such that
$\phi(x + \alpha p) - \phi(x) \leq -\ln(e/2)$ and $1 + \alpha q_j > 0$, $j = 0, 1, \ldots, n$.
A possible choice that enforces these conditions is $\alpha = 1/(1 + \|q\|_\infty)$.
In practice, selection of an approximate minimizer of $\phi(x + \alpha p)$ is recommended.
4. *Updating* $x^{k+1} := X_k(\mathbf{1} + \alpha q)$

$$\left(5.\ \textit{Scaling} \qquad x^{k+1} := \frac{1}{(x^{k+1})_0} x^{k+1}\right)$$

Return to 1 with $k := k + 1$.
End.

The infeasibility test (step 2.1) is based on the first least squares equation, $q + X_k A'u = \mathbf{1}$. It is a straightforward application of a theorem on alternatives (see [2]), that $A'u > 0$ implies that F is infeasible. The feasibility test (step 2.2) is based on the second least squares equation $AX_k q = 0$. It is also straightforward to validate.

We have put the scaling step in parentheses to stress that it need not be performed. Since the algorithm deals with a homogeneous problem, the sequence of rays in $\mathfrak{R}^{n+1}$ that it generates is independent of the positive scaling that is used. Actually, scaling the variable x^k to $(x^k)_0 = 1$ is to be performed only when one wants to interpret the vector x^k in terms of F and not in terms of the homogeneous problem $\tilde{F}$.

The convergence analysis is based on the lemma

Lemma 4.1. *Let* $x \in \operatorname{int} \mathfrak{R}_+^{n+1}$ *and* $q \in \mathfrak{R}_+^{n+1}$ *be such that*

$$AXq = 0, \tag{4.4}$$

$$\mathbf{1}'q = \|q\|^2, \tag{4.5}$$

$$q_{\max} = \max_j q_j \geq 1. \tag{4.6}$$

Then

$$\phi(X(\mathbf{1} + \alpha q)) - \phi(x) \leq -\ln\left(\frac{e}{2}\right) \tag{4.7}$$

for $\alpha = (1 + \|q\|_\infty)^{-1}$.

Any solution to (4.2) and (4.3) satisfies (4.4) and (4.5). If the infeasibility test has not been met (4.6) also holds. Lemma 4.1 applies. It shows that step 3 of the algorithm is well defined and that at each iteration the potential is decreased by at least the right-hand side of (4.7), a quantity that is independent of x. It follows that the algorithm converges in at most K iterations, where K is the smaller integer larger than $(\ln \varepsilon)/(1 - \ln 2)$. Upon choosing ε small enough with respect to the size of the problem, one can prove that the algorithm solves the feasibility problem in $O(nL)$ iterations. For a standard proof we refer to [3].

The overall complexity of the algorithm is determined by the computational effort per iteration. It consists essentially in solving the least squares equations (4.2) and (4.3), which in the dense case requires $O(m^2 n)$ operations. Several solution techniques are available for solving (4.2) and (4.3). The main approaches are the factorization of an appropriate matrix, on the one hand, and iterative techniques, on the other hand.

The more common factorizations are a Cholesky decomposition of AX^2A' and a QR decomposition of XA'. A Cholesky decomposition of AX^2A' can be used to solve the system $AX^2A'u = AX\mathbf{1}$, from which a solution to (4.2) and (4.3) is easily constructed. The QR decomposition of XA' can be used to compute the projection matrix P_{AX} on the null space of AX. Then $q = P_{AX}\mathbf{1}$. The drawback of these approaches is that it is rather difficult to update the decomposition from one iteration to the next, because at each iteration *all* the components of x change and thus *all* the columns of AX change. A full decomposition is to be performed at each iteration, which may be a prohibitively time-consuming task.

The iterative methods, such as the conjugate gradient method, are likely to deliver good approximate solutions very quickly, and thus at a low computational cost, provided some kind of preconditioning is used. However, at first sight, the arguments that are used in the convergence analysis and lead to polynomial complexity rely crucially on properties (4.4) and (4.5) of exact projections, which usually are lost when taking approximations.

The object of the next section is to show that a relaxation of the conditions in Lemma 4.1 makes it possible to accommodate approximate projections.

§3. A Relaxed Convergence Lemma

In order to relax the hypotheses of Lemma 4.1, we introduce three parameters, or tolerance levels, θ_1, θ_2, and θ_3.

Lemma 4.2. *Let θ_1, θ_2, and θ_3 be nonnegative parameters satisfying*

$$\theta_1 \geq 0, \theta_2 \geq 0, 1 \geq \theta_3 > 0, \quad \text{and}$$

$$(1 + \theta_3)^{-1}(\theta_1 + \theta_2) - (\theta_3 - \ln(1 + \theta_3)) = \delta < 0. \tag{4.8}$$

Let $x \in \operatorname{int} \Re_+^{n+1}$ and $q \in \Re_+^{n+1}$ be such that

$$\|AXq\| \leq \theta_1 (n + 1)^{-1} \|Ax\|, \tag{4.9}$$

$$\mathbf{1}'q \geq \|q\|^2 - \theta_2, \quad \textit{and} \tag{4.10}$$

$$q_{\max} \geq \theta_3. \tag{4.11}$$

Then

$$\phi(X(1 + \alpha q)) - \phi(x) \leq \delta \quad \textit{for } \alpha = (1 + \|q\|_\infty)^{-1}. \tag{4.12}$$

PROOF. Direct computations show that

$$\phi(X(\mathbf{1} + \alpha q)) - \phi(x) = (n + 1)\ln\frac{\|Ax + \alpha AXq\|}{\|Ax\|} - \sum_{j=0}^{n} \ln(1 + \alpha q_j). \tag{4.13}$$

In view of (4.10) one gets

$$(n + 1)\ln\frac{\|Ax + \alpha AXq\|}{\|Ax\|} \leq (n + 1)\ln\left(1 + \alpha\frac{\|AXq\|}{\|Ax\|}\right) \leq \theta_1(1 + \theta_3)^{-1}. \tag{4.14}$$

From the definition of α, one has $\alpha|q_j| < 1$. Expanding the logarithm yields

$$\ln(1 + \alpha q_j) = \alpha q_j - \sum_{k=0}^{\infty} \frac{(-\alpha q_j)^k}{k}.$$

Thus

$$\sum_{j=0}^{n} \ln(1 + \alpha q_j) \geq \alpha\mathbf{1}'q - \sum_{j=0}^{n}\sum_{k=2}^{\infty} \frac{(-\alpha q_j)^k}{k}.$$

For the sake of simpler notation we denote $t = \|q\|_\infty$. Since $|q_j| \leq t$, then, for $k \geq 2$, $|q_j/t|^k \leq (q_j/t)^2$. It follows that

$$\sum_{j=0}^{n} |q_j|^k \leq t^k \sum_{j=0}^{n} (q_j/t)^2 = t^k(\|q\|/t)^2.$$

The series $\sum_{k=2}^{\infty}((\alpha t)^k/k)$ converges. It is thus legitimate to interchange summations. Adding and subtracting the missing term $\alpha t^k(\|q\|/t)^2$ yields

$$\sum_{j=0}^{n} \ln(1 + \alpha q_j) \geq \alpha\mathbf{1}'q + (\|q\|/t)^2\left(\alpha t - \sum_{k=1}^{\infty}\frac{(\alpha t)^k}{k}\right)$$

$$\geq \alpha\mathbf{1}'q + (\|q\|/t)^2(\alpha t + \ln(1 - \alpha t)).$$

In view of (4.9) we obtain

$$\sum_{j=0}^{n} \ln(1 + \alpha q_j) \geq -\alpha\theta_2 + (\|q\|/t)^2(\alpha t(1 + t) + \ln(1 - \alpha t)).$$

Replacing α by its value yields

$$\sum_{j=0}^{n} \ln(1 + \alpha q_j) \geq -(1 + t)^{-1}\theta_2 + (\|q\|/t)^2(t - \ln(1 + t)).$$

The function $t - \ln(1 + t)$ is increasing. In view of (4.11) it achieves its minimum value at $t = \theta_3$. Hence

$$\sum_{j=0}^{n} \ln(1 + \alpha q_j) \geq -(1 + \theta_3)^{-1}\theta_2 + (\theta_3 - \ln(1 + \theta_3)). \tag{4.15}$$

(4.13) together with (4.14) and (4.15) yields (4.12).

If $\theta_1 = 0 = \theta_2$ and $\theta_3 = 1$, Lemma 4.2 is just the same as Lemma 4.1. Algorithm 4.1 is now modified to account for approximate projections.

Algorithm 4.2

Initialization. Let $\varepsilon > 0$ be a tolerance level. Let θ_1, θ_2, and θ_3 be nonnegative parameters satisfying condition (4.8) of Lemma 4.2. Let $x^0 = \mathbf{1} \in \mathfrak{R}^{n+1}$ and let $k = 0$.

Basic iteration. $x^k \in \operatorname{int} \mathfrak{R}_+^{n+1}$, with $(x^k)_0 = 1$, is given.

1. *Approximate projection step.* Find the $q \in \mathfrak{R}^{n+1}$ satisfying conditions (4.9), (4.10), and (4.11) of Lemma 4.2.
2. *Convergence tests.*
   2.1. (*infeasibility test*) If no q satisfies (4.9) to (4.11), STOP: F has no feasible solution.
   2.2. (ε-*convergence*) If $\exp\{\phi(x^k)\} < \varepsilon$ STOP: x^k is an ε-solution to F.
3. *Line search.* Find $\alpha \in \mathfrak{R}_+$ such that $\phi(x + \alpha p) - \phi(x) \leq \delta = (1 + \theta_3)^{-1}(\theta_1 + \theta_2) - (\theta_3 - \ln(1 + \theta_3))$ and $1 + \alpha q_j > 0, j = 0, 1, \ldots, n$.
4. *Updating* $x^{k+1} := X_k(\mathbf{1} + \alpha q)$

(5. *Scaling* $x^{k+1} := \dfrac{1}{(x^{k+1})_0} x^{k+1}$)

Return to 1 with $k := k + 1$.

End.

If $\theta_1 > 0$, the line search in step 3 of Algorithm 4.2 must account for the term $\ln \|AX(\mathbf{1} + \alpha q)\|$, which is no longer independent of α. As a result the one-dimensional problem in step 3 is no longer convex. Yet it remains simple enough. In particular, the derivatives of $\phi(x + \alpha p)$ with respect to α are easily computed.

Theorem 4.1. *Let θ_1, θ_2, and θ_3 be nonnegative parameters satisfying condition (4.8) of Lemma 4.2. Algorithm 4.2 converges in at most K iterations, where K is the smallest integer greater than (Log ε)/δ.*

PROOF. Let us first show that the conclusion in step 2.1 is valid. Let q be the exact projection. It satisfies (4.9) and (4.10) with $\theta_1 = 0 = \theta_2$. If the algorithm stops at step 2.1 one has $q_{\max} < \theta_3 \leq 1$. Hence F has no solution. If, on the contrary, the convergence tests are not met, then according to Lemma 4.2 the line search will yield a decrease of the potential of at least δ. Hence the result.

The closer to zero are ε and δ, the larger is K. However, ε and δ play a different role. The parameter ε must be chosen so as to ensure that $\exp\{\phi(x)\} < \varepsilon$ implies feasibility. Its value depends on the size L of the problem. On the contrary, δ is only user dependent. Its value is defined by the desired level of accuracy in solving the least squares equations. For a fixed value of $\delta < 0$, it is easily shown that that Theorem 4.1 implies that Algorithm 4.2 solves the feasibility problem in $O(nL)$ iterations. δ influences the parameters in the polynomial function but not the nature of the function itself.

We conclude this section with a word about linear programming problems with an unknown optimal value. Let $\text{Min}\{c'x \mid Bx = b, x \geq 0\}$ be the problem at hand. We paraphrase the approach in [3]. Let $D(z)$ be the matrix $(-b, B)$ augmented with the row $(-z, c')$, where z is a known lower bound of the problem. We consider the parametrized feasibility problem in $\mathfrak{R}^{n+1}$: $D(z)x = 0$, where z is the parameter in $\mathfrak{R}$. Let $(q(z), u(z))$ solve the least squares equations. If the infeasibility test is met then $D(z)'u(z) > 0$. The lower bound z can be increased until one of the components of $D(z)'u(z)$ takes the value zero. That suggests that computing the vector u in the least squares equation (or an approximation of it) may be worthwhile. This would be an approach very similar to the one in [8].

§4. Approximate Projections

The exact projection must satisfy the two least squares equations

$$q + XA'u = \mathbf{1}, \tag{4.2}$$

$$AXq = 0. \tag{4.3}$$

We see two ways of getting approximate solutions to (4.2) and (4.3), just by forcing one of the two equations to be solved exactly and the other one approximately. Following [8], we shall speak of an approximation in the range space if (4.2) is satisfied and (4.3) is approximated. If (4.3) is satisfied and (4.2) is approximated we shall speak of an approximation in the null space.

4.1. Approximations in the Range Space

Let θ_1, θ_2, and θ_3 be nonnegative parameters satisfying condition (4.8) of Lemma 4.2. Assume a CG (conjugate gradient) method is used to solve

$$\text{Minimize } \tfrac{1}{2}\|XA'u - \mathbf{1}\|^2. \tag{4.16}$$

Let u be a solution of (4.16) and define q by $q = \mathbf{1} - XA'u$. If u is an exact solution of (4.16) then q is the exact projection. Otherwise q is an approximation in the range space. The CG method typically computes the gradient $AX(XA'u - \mathbf{1})$ of $\tfrac{1}{2}\|XA'u - \mathbf{1}\|^2$. This gradient is just the residual $r(u)$ of the equation

$$AX^2A'u - AX\mathbf{1} = 0. \tag{4.17}$$

Suppose the CG method is stopped before a solution of (4.17) has been found. Then $r(u) \neq 0$. Since $r(u) = AX(XA'u - \mathbf{1}) = -AXq$, condition (4.9) of Lemma 4.2 amounts to

$$\|r(u)\| \leq \frac{\theta_1}{n+1}\|r(0)\|. \tag{4.18}$$

Since $\|q\|^2 - \mathbf{1}'q = q'(XA'u) = -u'r(u)$, condition (4.10) is just

$$u'r(u) \geq -\theta_2. \tag{4.19}$$

Finally, since $\max_j q_j = 1 - \min_j (XA'u)_j$, condition (4.11) amounts to

$$\min_j (XA'u)_j \leq 1 - \theta_3. \tag{4.20}$$

Therefore a possible way to construct an approximation in the range space is to run a CG method in (4.16). In order to incorporate it in Algorithm 4.2, the CG method must be run until conditions (4.18), (4.19), and (4.20) are met. If the CG method terminates with an exact solution then $r(u) = 0$. If (4.20) is not met one concludes that F is infeasible.

4.2. Approximations in the Null Space

Assume that a basis of the null space of A is at hand. More specifically, let N be an $(n + 1 - m) \times (n + 1)$ matrix whose rows generate the null space of A and let v be in $\Re^{n+1-m}$. Thus any p satisfying $Ap = 0$ can be written $p = N'v$. Substituting $N'v$ for p in problem QP of Section 2 yields the unconstrained problem

$$\text{Minimize } \tfrac{1}{2}\|X^{-1}N'v - \mathbf{1}\|^2. \tag{4.21}$$

If the CG method is applied to solving (4.21) it typically computes the gradient $NX^{-1}(X^{-1}N'v - \mathbf{1})$ of $\frac{1}{2}\|X^{-1}N'v - \mathbf{1}\|^2$. This gradient is just the residual $s(v)$ of the equation

$$NX^{-2}N'v - NX^{-1}\mathbf{1} = 0. \tag{4.22}$$

Suppose the CG method is stopped before a solution of (4.22) has been found. Let $q = X^{-1}N'v$. By construction $AXq = AN'v = 0$. Condition (4.9) of Lemma 4.2 is satisfied with $\theta_1 = 0$. Since $\|q\|^2 - \mathbf{1}'q = v'NX^{-1}(X^{-1}N'v - \mathbf{1}) = v's(v)$, condition (4.10) amounts to

$$v's(v) \leq \theta_2. \tag{4.23}$$

Finally condition (4.11) amounts to

$$\max_j (X^{-1}N'v)_j \geq \theta_3. \tag{4.24}$$

Therefore a possible way to construct an approximation in the range space is to run a CG method on (4.21). In order to incorporate it in Algorithm 4.2, the CG method must be run until conditions (4.22), (4.23), and (4.24) are met. If the CG method terminates with an exact solution then $s(v) = 0$. If (4.24) is not met one concludes that F is infeasible.

The matrix N is not directly available. It must be constructed. We briefly discuss this point. A first possibility is to partition A in $A = (A_1, A_2)$, where A_1 is an $m \times m$ matrix with known inverse (a "basis" in standard linear programming theory.) Then $N = (-A_2'(A_1')^{-1}, I)$ is a generator of the null space of A. This approach is equivalent to working in a reduced space. Another possibility is to use a matrix of orthonormal vectors as a generator of the null space of A. Such a matrix can be obtained using, for instance, a QR factorization of A'. Let $(R, 0)Q$ be this factorization of A, where R is an $m \times m$ lower triangular matrix and $Q = \begin{bmatrix} Q_1 \\ Q_2 \end{bmatrix}$ is an $(n + 1) \times (n + 1)$ orthogonal matrix. The rows of Q_2 generate the null space of A. Note also that the projection matrix P_A corresponding to the null space of A is $Q_2'Q_2 = I - Q_1'Q_1$. The merit of this approach is that the time-consuming task of computing P_A (or equivalently Q_1) has to be done once and for all.

Let us show that it is possible to construct an infeasibility test from the search direction p that is delivered by the iterative scheme, for instance, the CG method, even though the direction is an approximate solution to the least squares problem. If p is an approximate solution, that is, $q = X^{-1}p$ is an approximate projection, it is not necessarily true that $\mathbf{1} - q = XA'u$ for some u. Thus we cannot conclude from $q < \mathbf{1}$ that the problem is infeasible. Since the projection matrix $Q_1'Q_1$ corresponding to the range space of A' is available, we suggest computing the closest vector to $X^{-1}(\mathbf{1} - q)$ in the range space of A', namely $Q_1'Q_1X^{-1}(\mathbf{1} - X^{-1}p)$. The test $Q_1'Q_1X^{-1}(\mathbf{1} - X^{-1}p) > 0$ is a valid infeasibility test. If p is the exact solution of the least squares problem, this test may not coincide with the standard infeasibility test $q < \mathbf{1}$. Yet it can be added to step 2.1 of Algorithm 4.2 for increased efficiency.

We conclude this section by mentioning an alternative way of constructing approximations in the null space. One can use directly a generator $\bar{N}$ of the null space of AX and solve

$$\text{Minimize } \tfrac{1}{2}\|\bar{N}'v - \mathbf{1}\|^2. \tag{4.25}$$

If $\bar{N}$ is an orthogonal basis the solution is immediate and there is no approximation problem. Of course, such a basis is not available in general. However, in some problems it is possible to partition AX in (A_1X_1, A_2X_2), where A_1X_1 is an $m \times m$ matrix with known inverse. Then one works in a reduced space and approximate projections can be computed in a manner quite similar to the one discussed above. This is the approach of [7].

§5. Partial Update of the Null Space Basis

Assume that we apply some method based on the previous section. As the algorithm progresses, the discrepancies between the coordinates of the current iterate x are increasing. Some of the variables go to zero very fast, whereas some others stay rather stable. As a result, the problem QP is likely to become more and more ill-conditioned.

We now present a modification, based on Lemma 4.2, which may help overcome this difficulty. As in Karmarkar's modified algorithm [12], it uses a partial updating of the working matrix AX. Let $X = X_a X_b X_c$ be a decomposition of X such that (i) X_a, X_b, and X_c are diagonal matrices with strictly positive entries, (ii) the QR decomposition of AX_a is known, (iii) the QR decomposition of AX_aX_b can be updated from the QR decomposition of AX_a, and (iv) X_c is not far from the identity matrix, in the sense that $\max\{\|X_c\|, \|X_c^{-1}\|\} \leq 1/\rho$, where ρ is a parameter close enough to 1.

Let us denote by $\bar{A}$ the matrix AX_aX_b.

Theorem 4.2. *Let* $0 < \rho < 1$ *be such that*

$$1 - \rho^2 - (1 + \rho^2)(\rho^2 - \ln(1 + \rho^2)) < 0. \tag{4.26}$$

Let X_c *satisfy* $\max\{\|X_c\|, \|X_c^{-1}\|\} \leq 1/\rho$. *The search direction* $q = -X_c^{-1}p$, *with* $p = P_{\bar{A}}X_c^{-1}\mathbf{1}$, *satisfies the hypotheses of Lemma* 4.2, *with* $\theta_1 = 0$, $\theta_2 = 1 - \rho^2$, *and* $\theta_3 = \rho^2$, *whenever F is feasible.*

Proof. From the definition of q there is a vector u in $\mathfrak{R}^m$ such that $p = X_c^{-1}\mathbf{1} - A'u$. The infeasibility test $A'u > 0$ amounts to $X_c^{-1}\mathbf{1} - p > 0$. Assume F is feasible. Then $\max_j(p_j - 1/x_{cj}) \geq 0$. Since $\max_j p_j - \rho \geq \max_j(p_j - 1/x_{cj})$, it amounts to $\max_j p_j \geq \rho$. Thus $\|p\| \geq \rho$. Let $q = X_c^{-1}p$; then $\max_j q_j \geq \rho^2$. (11) holds with $\theta_3 = \rho^2$.

Let us bound the quantity

$$\mathbf{1}'q = \mathbf{1}'X_c^{-1}P_{\bar{A}}X_c^{-1}\mathbf{1} = \|P_{\bar{A}}X_c^{-1}\mathbf{1}\|^2 = \|p\|^2.$$

Note that $\|q\| \leq \|X_c^{-1}\|\,\|p\|$. Hence, $\mathbf{1}'q - \|q\|^2 \geq \|p\|^2(1 - 1/\rho^2) \geq -(1 - \rho^2)$, from which it follows that $\mathbf{1}'q \geq \|q\|^2 - (1 - \rho^2)$. Hence (4.10) holds with $\theta_2 = 1 - \rho^2$. Finally $AXq = \bar{A}X_c p = \bar{A}p = 0$. (4.9) holds with $\theta_1 = 0$. The theorem is proved.

It is easily verified that the first step of the conjugate gradient scheme applied to

$$\text{Minimize } \{\tfrac{1}{2}\|X_c^{-1}p - \mathbf{1}\|^2 | (AX_aX_b)p = 0\}$$

generates the same search direction $q = X_c^{-1}p$ as in Theorem 4.2.

As far as practical implementation is concerned, one could couple partial updating with one, or a few, iterations of the conjugate gradient method. We now describe this mixed procedure.

Let x be the current iterate and let $x_a \in \operatorname{int} \mathfrak{R}_+^{n+1}$ be an approximation of it such that an orthogonal basis of the null space of AX_a is known (or, equivalently, such that the projection matrix corresponding to the null space of AX_a is available). Let x_b and $x_c \in \operatorname{int} \mathfrak{R}_+^{n+1}$ be such that $X = X_a X_b X_c$. The decomposition is arbitrary but it must satisfy $\max\{\|X_c\|, \|X_c^{-1}\|\} \leq 1/\rho$. Moreover, there should be as many as possible components of x_b equal to 1. In the procedure an orthogonal basis of the null space of $AX_a X_b$ is computed. Then a CG scheme similar to the one given in Section 4 is used to solve approximately

$$\text{Minimize } \{\tfrac{1}{2}\|X_c^{-1}p - \mathbf{1}\|^2 | (AX_a X_b)p = 0\}.$$

The updating of the orthogonal basis amounts to the updating of the QR decomposition of AX_a after successive rank-one corrections. Indeed, if X_b has only one diagonal component different from 1, $AX_a X_b$ differs from AX_a by a single column, which is equivalent to a rank-one correction. The updating of the QR decomposition is $O(mn)$ in the dense case; see [9]. The number of rank-one corrections is equal to the number of coefficients in x_b which are different from 1.

It seems that this number is unambiguously determined by ρ. This is not quite so. Recall that the scaling step in Algorithm 4.2 is arbitrary. One may try different scaling in order to maximize the number of indices j such that x_j/x_{aj} is in the interval $[\rho, 1/\rho]$. A possible choice is to scale x to $\tilde{x} = \sigma^{-1}x$, where $\sigma = [\prod_{j=0}^{n}(x_{aj}/x_j)]^{1/(n+1)}$. Then the geometric mean $[\prod_{j=0}^{n}(\tilde{x}_j/x_{aj})]^{1/(n+1)}$ of the critical numbers $\tilde{x}_j/x_{aj}$ takes the value 1. Finally, we note that if X_c is the identity matrix, the search direction is the same as in Algorithm 4.1.

Acknowledgment

The author would like to thank Yves Pochet for valuable discussions on an early draft of the paper. He also thanks an anonymous referee and C. Fraley for their constructive comments on a first version. This work was initiated while the author was still at CORE. CORE is kindly acknowledged for its support and its stimulating environment.

References

[1] I. Adler, M. C. G. Resende, and G. Veiga, An implementation of Karmarkar's algorithm for linear programming, Working Paper, Operations Research Center, University of California, Berkeley (1986).

[2] G. de Ghellinck and J.-P. Vial, An extension of Karmarkar's algorithm for solving a system of linear homogeneous equations on the simplex, *Math. Programming* **39** (1987), 79–92.

[3] G. de Ghellinck and J.-P. Vial, A polynomial Newton method for linear programming, *Algorithmica* **1** (1986), 425–453.

[4] R. S. Dembo and T. Steihaug, Truncated-Newton algorithms for large-scale unconstrained optimization, *Math. Programming* **26** (1983) 190–212.

[5] J. E. Dennis, A. M. Morshedi, and K. Turner, A variable metric variant of Karmarkar's algorithm for linear programming, *Math. Programming* **39** (1987), 1–20.

[6] P. E. Gill, W. Murray, M. A. Saunders, J. A. Tomlin, and M. H. Wright, On projected Newton barrier methods for linear programming and an equivalence to Karmarkar's projective method, *Math. Programming* **36** (1986), 183–209.

[7] D. Goldfarb and S. Mehrotra, A relaxed version of Karmarkar's method, *Math. Programming* **40** (1988), 289–315.

[8] D. Goldfarb and S. Mehrotra, Relaxed variants of Karmarkar's algorithm for linear programs with unknown optimal objective value, *Math. Programming* **40** (1988), 183–195.

[9] G. H. Golub and C. Van Loan, *Matrix Computations*, John Hopkins University Press, Baltimore (1983).

[10] C. Gonzaga, Conical projections algorithms for linear programming, Memorandum N. UCB/ERL M85/61, revised version UCB/ERL M87/11, Electronics Research Laboratory, College of Engineering, University of California, Berkeley (1985, revised 1987).

[11] M. T. Heath, Numerical methods for large sparse linear least squares problems, *SIAM J. Sci. Stat. Comput.* **5** (1984), 562–589.

[12] N. Karmarkar, A new polynomial time algorithm for linear programming, *Combinatorica* **4**(4) (1984), 373–395.

[13] I. J. Lustig, A practical approach to Karmarkar's algorithm, SOL 85-5, Systems Optimization Laboratories, Department of Operations Research, Stanford University, Stanford, Calif. (1985).

[14] M. Minoux, New suggested implementations of Karmarkar's algorithm, Cahier n°.71, Laboratoire d'Analyse et Modélisation de Systèmes pour l'Aide à la Décision, Université de Paris-Dauphine (1986).

[15] C. C. Paige and M. A. Saunders, LSQR: An algorithm for sparse linear equations and sparse least squares, *ACM Trans. Math. Software* **8**(1) (1982), 43–71.

[16] P. F. Pickel, Approximate projections for the Karmarkar algorithm, Manuscript, Polytechnic Institute of New York, Farmingdale, N.Y. (1985).

[17] D. Shanno, Computing Karmarkar projections quickly, Working Paper 85-10, *Math. Programming* **41** (1988), 61–71.

[18] D. Shanno and R. Marsten, On implementing Karmarkar's method, Working Paper 85-01, Graduate School of Business Administration, University of California, Davis (1985).

[19] T. Steihaug, The conjugate gradient method and trust regions in large scale optimization, Technical Report 81-1, Department of Mathematical Sciences, Rice University, Houston (1981).

[20] M. J. Todd and B. P. Burell, An extension of Karmarkar's algorithm for linear programming using dual variables, *Algorithmica* **1** (1986), 409–424.

[21] P. Toint, Towards an efficient sparsity exploiting Newton method for minimization, *Sparse Matrices and Their Uses*, I. Duff (ed.) Academic Press, New York, 1981.

CHAPTER 5

A Locally Well-Behaved Potential Function and a Simple Newton-Type Method for Finding the Center of a Polytope

Pravin M. Vaidya

Abstract. The center of a bounded full-dimensional polytope $P = \{x: Ax \geq b\}$ is the unique point ω that maximizes the strictly concave potential function $F(x) = \sum_{i=1}^{m} \ln(a_i^T x - b_i)$ over the interior of P. Let x_0 be a point in the interior of P. We show that the first two terms in the power series of $F(x)$ at x_0 serve as a good approximation to $F(x)$ in a suitable ellipsoid around x_0 and that minimizing the first-order (linear) term in the power series over this ellipsoid increases $F(x)$ by a fixed additive constant as long as x_0 is not too close to the center ω.

§1. Introduction

Let the polytope P be defined as

$$P = \{x: Ax \geq b\}$$

where $x \in R^n$, $b \in R^m$, and $A \in R^{m \times n}$. The center ω of the polytope P is defined to be the unique point that maximizes the strictly concave potential function

$$F(x) = \sum_{i=1}^{m} \ln(a_i^T x - b_i).$$

Under the assumption that the polytope P is bounded and has a nonzero interior, the Hessian of F is negative definite over the interior of P, and so ω is indeed a unique point. We let $f(x) = F(\omega) - F(x)$ denote the normalized potential corresponding to $F(x)$. We define transformed coordinates $\Psi_i(x)$ as

$$\Psi_i(x) = \frac{a_i^T(x - \omega)}{a_i^T \omega - b_i}, \qquad i = 1, 2, \ldots, m.$$

The coordinates $\Psi_i(x)$ were originally defined in [3, 4] and will be used extensively here in proving various properties of $f(x)$. Let $\Sigma(\delta)$ denote the ellipsoid around ω given by

$$\Sigma(\delta) = \left\{x: \sum_{i=1}^{m} \Psi_i(x)^2 \leq \delta^2\right\}.$$

Then $\Sigma(1) \subseteq P \subseteq \Sigma(m)$ [3,4]. Thus the ratio of the maximum to the minimum distance from the center ω to any point on the boundary of P is upper bounded by m. So the center is a balanced point in the polytope. The center plays an important role in algorithms for linear and convex programming [1–6]. In particular, a polynomial time algorithm for computing a good approximation to the center can be converted into a polynomial time algorithm for linear programming [3,6].

In this short chapter we shall describe some properties of $f(x)$ and see how they may be used to develop an algorithm for finding the center ω. We shall assume that an initial point strictly in the interior of the polytope is available.

§2. Local Behavior of the Potential

Let x_0 be a point in the interior of the polytope P, let η be the gradient of $f(x)$ evaluated at x_0, and let H denote the Hessian of $f(x)$ evaluated at x_0. Explicitly,

$$\eta = -\sum_{i=1}^{m} \frac{1}{(a_i^T x_0 - b_i)} a_i$$

and

$$H = \sum_{i=1}^{m} \frac{1}{(a_i^T x_0 - b_i)^2} a_i a_i^T.$$

Let $E(r)$ be the ellipsoid around x_0 defined as

$$E(r) = \{x: (x - x_0)^T H (x - x_0) \leq r^2\}.$$

Note that if $0 \leq r \leq 1$ then the ellipsoid $E(r)$ is contained within the polytope P. We shall show that minimizing the linear function $\eta^T x$ over the ellipsoid $E(r)$ gives a good reduction in $f(x)$. Specifically, $f(x)$ is reduced by at least a fixed additive constant if x_0 lies outside the ellipsoid $\Sigma(0.5)$, whereas $f(x)$ is reduced by at least a fixed fraction if x_0 is within the ellipsoid $\Sigma(0.5)$.

Let $x = x_0 + t\xi$. Using the power series expansion of $f(x_0 + t\xi)$ at x_0, $f(x_0 + t\xi)$ may be written as

$$f(x_0 + t\xi) = f(x_0) + t\eta^T \xi + \frac{t^2}{2} \xi^T H \xi + \sum_{j=3}^{\infty} \frac{(-1)^j t^j}{j} \left(\sum_{i=1}^{m} \frac{(a_i^T \xi)^j}{(a_i^T x_0 - b_i)^j} \right).$$

We shall prove the following lemmas.

Lemma 5.1. *Let r be a parameter such that $0 \leq r < 1$, and let $x = x_0 + t\xi$ be a point in the ellipsoid $E(r)$ around x_0. Then*

$$\frac{t^2}{2}\xi^T H \xi \le 0.5r^2$$

and

$$\left|\sum_{j=3}^{\infty}\frac{(-1)^j t^j}{j}\left(\sum_{i=1}^{m}\frac{(a_i^T\xi)^j}{(a_i^T x_0 - b_i)^j}\right)\right| \le \frac{r^3}{3(1-r)}.$$

The next lemma lower bounds the maximum change obtainable in the linear function $\eta^T x$ over the ellipsoid $E(r)$.

Lemma 5.2. *Let r be a parameter such that $0 \le r < 1$, and let δ be a parameter such that $0 \le \delta < 1$. Let x' be the point where the straight line joining x_0 to the center ω intersects the boundary of the ellipsoid $E(r)$ around x_0. The point x' satisfies the following conditions.*

(1) *If $x_0 \notin \Sigma(\delta)$ then $\eta^T(x_0 - x') \ge \dfrac{\delta(1-\delta)}{2(1+\delta)}r$.*

(2) *If $x_0 \in \Sigma(\delta)$ then $\eta^T(x_0 - x') \ge ((1-\delta)f(x_0))^{1/2} r$.*

The final lemma says that the point that minimizes the linear function $\eta^T x$ over $E(r)$ gives a good reduction in $f(x)$.

Lemma 5.3. *Let δ be a parameter such that $0 < \delta < 0.7$, and let ε be a parameter such that $0 < \varepsilon < 1$. Let r_0 be defined by*

$$r_0 = \begin{cases} \varepsilon, & \text{if } x_0 \notin \Sigma(\delta) \\ \varepsilon\sqrt{f(x_0)}, & \text{if } x_0 \in \Sigma(\delta) \end{cases}.$$

Let $\mathbf{x}$ be the point that minimizes the linear function $\eta^T x$ over the ellipsoid $E(r_0)$ around x_0. *Then point $\mathbf{x}$ satisfies the following conditions.*

(1) *If $x_0 \notin \Sigma(\delta)$ then*

$$f(\mathbf{x}) - f(x_0) \le -\frac{\delta(1-\delta)}{2(1+\delta)}\varepsilon + 0.5\varepsilon^2 + \frac{\varepsilon^3}{3(1-\varepsilon)}$$

(2) *If $x_0 \in \Sigma(\delta)$ then*

$$f(\mathbf{x}) \le \left(1 - \varepsilon(1-\delta)^{1/2} + 0.5\varepsilon^2 + \frac{\varepsilon^3}{3(1-\varepsilon)}\right) f(x_0).$$

Lemma 5.3 may be converted into an algorithm for finding the center ω as follows. Let x_0 be the current point and let $\mathbf{x}$ be the point that minimizes $\eta^T x$ over the ellipsoid $E(r_0)$, where r_0 is as defined in Lemma 5.3. $x_0 - \mathbf{x}$ satisfies the system of linear equations

$$H(x_0 - \mathbf{x}) = \mathbf{t}\eta$$

for some scalar $\mathbf{t}$. So we may compute a direction ξ by solving the system

$$H\xi = \eta$$

and minimize $f(x)$ on the line $x_0 + t\xi$. (Note that the one-dimensional minimization need not be exact.) According to Lemma 5.3, the point thus obtained will reduce $f(x)$ by an additive constant if $x_0 \notin \Sigma(0.5)$ and will reduce $f(x)$ by a fixed fraction if $x_0 \in \Sigma(0.5)$. Thus starting with an initial point x_{init} strictly in the interior of the polytope P we can produce a sequence of points converging to the center ω.

We now give a more formal algorithm for computing the center. We shall assume that a point x_{init} such that $f(x_{\text{init}}) \leq M$ is available. (x_{init} is in the interior of polytope P.) The algorithm produces a sequence of points $z_0 = x_{\text{init}}$, $z_1, \ldots, z_k, \ldots$ that converge to the center ω. x_0 denotes the current point in the computation, and k is the step number. Let θ be a parameter less than 1/250. The output of the the algorithm is a point z_q such that $f(z_q) \leq \theta$.

Algorithm Find-Center
$x_0 := x_{\text{init}}$; $z_0 := x_{\text{init}}$; $k := 0$;
Loop:
  /*x_0 **is the current point** */
  Let η be the gradient of $f(x)$ at x_0 and
  let H be the Hessian of $f(x)$ at x_0;
  Let ξ be the direction obtained by solving $H\xi = \eta$;
  /* **Increment step number** k **and compute** z_k */
  $k := k + 1$;
  Let z_k be the point that minimizes $f(x)$ on the line
  $x_0 + t\xi$ where t is a scalar;
  /* **Reset** x_0 */
  $x_0 := z_k$;
  If $f(z_k - 1) - f(z_k) \geq \theta/3$ then go to Loop
  else halt;
end Find-Center

It is worth noting that ξ is in the same direction as the direction generated by the Newton-Raphson method applied to the problem of minimizing $f(x)$; however, the length of the step taken in the direction of ξ is quite different in the above algorithm. Also note that in the above algorithm the value of F at ω is not required to compute the difference $f(z_{k-1}) - f(z_k)$. Furthermore, the one- dimensional minimization on the line $x_0 + t\xi$ need not be performed exactly. (It suffices to find a point on this line where f is either reduced by a fixed additive constant or f is reduced by a fixed fraction.)

Let z_q be the point at the termination of the algorithm. Then $f(z_{q-2}) \geq \theta/3$, and $f(z_q) \leq \theta$. That $f(z_{q-2}) \geq \theta/3$ follows from the observations that the algorithm did not terminate at the $(q-1)$st step and that the minimum value of f is zero. That $f(z_q) \leq \theta$ is shown as follows. We have that

$$f(z_{q-1}) - f(z_q) < \theta/3,$$

and hence by Lemma 5.3 (with $\delta = 1/2$, $\varepsilon = 1/10$), z_{q-1} cannot be outside

$\Sigma(0.5)$. Therefore $z_{q-1} \in \Sigma(0.5)$. Then again by Lemma 5.3 (with $\delta = 1/2$, $\varepsilon = 0.5$) we get that

$$f(z_{q-1}) - f(z_q) \geq \tfrac{1}{3} f(z_{q-1}).$$

Thus $f(z_q) \leq f(z_{q-1}) \leq \theta$.

We shall now show an upper bound of $O((mn^2 + n^3)(M + \log(1/\theta)))$ on the total number of arithmetic operations performed by the algorithm. Since during a step we must compute the gradient η and the Hessian H and solve a system of linear equations, executing a step requires $O(mn^2 + n^3)$ arithmetic operations. So to obtain the said bound on the total number of arithmetic operations it suffices to show a bound of $O(M + \log(1/\theta))$ on the total number of steps executed by the algorithm. The number of steps is upper bounded as follows. It requires $O(M)$ steps to obtain a point in $\Sigma(0.3)$ because by Lemma 5.3 $f(x)$ is decreased by a fixed additive constant at each step as long as the current point is outside $\Sigma(0.3)$. Let p be the step number such that $z_p \in \Sigma(0.3)$ and for $0 \leq k < p$, $z_k \notin \Sigma(0.3)$. By Lemma 5.5 in Section 3, we get that $f(z_p) \leq 0.1$ and that for all $k > p$, $f(z_k) \leq f(z_p) \leq 0.1$. Then by Lemma 5.6 in Section 3 it follows that $z_k \in \Sigma(0.5)$ for all $k > p$. So we can apply Lemma 5.3 to each step after the pth step and conclude that from the pth step onward $f(x)$ decreases by a fixed fraction at each step. Hence in $O(\log(1/\theta))$ steps after the pth step $f(x)$ must fall below $\theta/3$. Thus the total number of steps is $O(M + \log(1/\theta))$.

In a manner similar to [2, 6] it is possible to reduce the time complexity of the above algorithm by using an approximate Hessian H_a at each step where

$$H_a = \sum_{i=1}^{m} \frac{\Delta_i}{(a_i^T x_0 - b_i)^2} a_i a_i^T$$

and $\Delta_i \in [1/1.1, 1.1]$, $1 \leq i \leq m$. The above algorithm is modified as follows. The direction ξ is now computed by solving the system $H_a \xi = \eta$ instead of the system $H\xi = \eta$. The approximate Hessian H_a and its inverse are maintainedby performing rank-one updates as described in [2, 6]. z_k is still obtained by a one-dimensional minimization on the line $x_0 + t\xi$. We note that the direction ξ computed by the modified algorithm is now quite different from the Newton-Raphson direction. The modified algorithm still requires $O(M + \log(1/\theta))$ steps, which may be seen as follows. Suppose $E_a(r)$ is the ellipsoid defined as

$$E_a(r) = \{x: (x - x_0)^T H_a (x - x_0) \leq r^2\}.$$

Then

$$E\left(\frac{r}{1.1}\right) \subseteq E_a(r) \subseteq E(1.1r).$$

So Lemmas 5.1 through 5.3 are still valid (but with different constants) if the ellipsoid $E(r)$ is replaced by the ellipsoid $E_a(r)$. Thus minimizing $\eta^T x$ over $E_a(r)$ rather than $E(r)$ still gives an adequate decrease in $f(x)$. Use of approximate Hessians reduces the average work per step to $O(mn + m^{0.5} n^2)$ arithmetic operations, leading to a total of $O((mn + m^{0.5}n^2)(M + \log(1/\theta)))$ arithmetic operations.

§3. Proofs of Lemmas

We shall first prove a couple of lemmas that will be used in the proofs of the lemmas stated in Section 2.

Lemma 5.4. *Let $\Psi_i(x) = (a_i^T x - a_i^T \omega)/(a_i^T \omega - b_i)$, for $i = 1, 2, \ldots, m$. Then for any point x in the polytope P,*

$$\sum_{i=1}^{m} \Psi_i(x) = 0.$$

PROOF. A proof is given in [3, 4] but we include it here for completeness. Since the gradient of $f(x)$ vanishes at ω, taking the dot product of the gradient of $f(x)$ at ω with $\omega - x$ gives

$$-\sum_{i=1}^{m} \frac{a_i^T(\omega - x)}{(a_i{}^T \omega - b_i)} = \sum_{i=1}^{m} \Psi_i(x) = 0.$$

The next lemma bounds for the maximum value of $f(x)$ in the region $\Sigma(\delta)$ for $0 \leq \delta < 1$.

Lemma 5.5. *Let δ be a parameter such that $0 \leq \delta < 1$. Then the maximum value of $f(x)$ in the ellipsoid $\Sigma(\delta)$ is at most $\delta^2/2(1 - \delta)$.*

PROOF. Since $f(x)$ is strictly convex, and ω minimizes $f(x)$, the maximum value of $f(x)$ over the region $\Sigma(\delta)$ is achieved on the boundary of $\Sigma(\delta)$. We have

$$f(x) = \sum_{i=1}^{m} \ln\left(\frac{1}{1 + \Psi_i(x)}\right).$$

Using the Taylor series expansion, on the boundary of $\Sigma(\delta)$ we may write $f(x)$ as

$$f(x) = \sum_{j=1}^{\infty} \sum_{i=1}^{m} \frac{(-1)^j \Psi_i(x)^j}{j}$$

From Lemma 5.4, $\sum_{i=1} \Psi_i(x) = 0$. So on the boundary of $\Sigma(\delta)$ we get

$$\begin{aligned} f(x) &= \sum_{j=2}^{\infty} \sum_{i=1}^{m} \frac{(-1)^j \Psi_i(x)^j}{j} \\ &\leq \sum_{i=1}^{m} \frac{\Psi_i(x)^2}{2}\left(1 + \frac{2\delta}{3} + \frac{2\delta^2}{4} + \frac{2\delta^3}{5} + \cdots\right) \\ &\leq \frac{\delta^2}{2(1 - \delta)}. \end{aligned}$$

The next lemma lower bounds the value of the function $\sum_{i=1}^{m}(1/(1+\Psi_i(x))-1)$ over the region $\{x: x \in P, x \notin \Sigma(\delta)\}$ for $0<\delta<1$.

Lemma 5.6. *Let δ be a parameter such that $0<\delta<1$. The minimum value of $\sum_{i=1}^{m}(1/(1+\Psi_i(x))-1)$ over the region $\{x: x \in P, x \notin \Sigma(\delta)\}$ is greater than or equal to $\delta^2/(1+\delta)$.*

PROOF. Let $g(x)=\sum_{i=1}^{m}(1/(1+\Psi_i(x))-1)$. By Lemma 5.4, $\sum_{i=1}^{m}\Psi_i(x)=0$, and so

$$g(x)=\sum_{i=1}^{m}\left(\frac{1}{1+\Psi_i(x)}+\Psi_i(x)-1\right)=\sum_{i=1}^{m}\frac{\Psi_i(x)^2}{1+\Psi_i(x)}.$$

The Hessian of $g(x)$ evaluated at a point x is the matrix $A^T D(x) A$ where $D(x)$ is a diagonal matrix whose ith diagonal entry $D_{ii}(x)$ is given by $D_{ii}(x)=2(a_i^T\omega-b_i)/(a_i^T x-b_i)^3$. As $A^T D(x) A$ is positive definite in the interior of P, $g(x)$ is strictly convex in the interior of P. Furthermore, the minimum value of $g(x)$ over the interior of P occurs at the center ω. Thus the minimum value of $g(x)$ over the region $\{x: x \in P, x \notin \Sigma(\delta)\}$ occurs on the boundary of $\Sigma(\delta)$. And on the boundary of $\Sigma(\delta)$,

$$g(x)=\sum_{i=1}^{m}\frac{\Psi_i(x)^2}{1+\Psi_i(x)}\geq\sum_{i=1}^{m}\frac{\Psi_i(x)^2}{1+\delta}\geq\frac{\delta^2}{1+\delta}.$$

We shall now prove the lemmas stated in Section 2. We shall restate the lemmas for convenience.

Lemma 5.1. *Let r be a parameter such that $0 \leq r<1$, and let $x=x_0+t\xi$ be a point in the ellipsoid $E(r)$ around x_0. Then*

$$\frac{t^2}{2}\xi^T H \xi \leq 0.5 r^2$$

and

$$\left|\sum_{j=3}^{\infty}\frac{(-1)^j t^j}{j}\left(\sum_{i=1}^{m}\frac{(a_i^T\xi)^j}{(a_i^T x_0-b_i)^j}\right)\right| \leq \frac{r^3}{3(1-r)}.$$

PROOF. By definition of $E(r)$,

$$t^2\xi^T H \xi \leq r^2.$$

As H may be written as

$$H=\sum_{i=1}^{m}\frac{1}{(a_i^T x_0-b_i)^2}a_i a_i^T$$

we get

$$t^2\xi^T H\xi = t^2\sum_{i=1}^{m}\frac{(a_i^T\xi)^2}{(a_i^T x_0-b_i)^2}\leq r^2.$$

Thus

$$\left|\sum_{j=3}^{\infty}\frac{(-1)^j t^j}{j}\left(\sum_{i=1}^{m}\frac{(a_i^T\xi)^j}{(a_i^T x_0 - b_i)^j}\right)\right| \le \sum_{j=3}^{\infty}\frac{r^j}{j}$$
$$\le \frac{r^3}{3(1-r)}.$$

Lemma 5.2. *Let r be a parameter such that $0 \le r < 1$, and let δ be a parameter such that $0 \le \delta < 1$. Let x' be the point where the straight line joining x_0 to the center ω intersects the boundary of the ellipsoid $E(r)$* around x_0. *The point x' satisfies the following conditions.*

(1) *If $x_0 \notin \Sigma(\delta)$ then* $\eta^T(x_0 - x') \ge \dfrac{\delta(1-\delta)}{2(1+\delta)}r.$

(2) *If $x_0 \in \Sigma(\delta)$ then* $\eta^T(x_0 - x') \ge ((1-\delta)f(x_0))^{1/2}r.$

PROOF. Let $x_0 - x' = \lambda u$ where u is the unit vector in the direction of $x_0 - x'$ and $\lambda = \|x_0 - x'\|_2$. Then

$$\lambda^2 u^T H u \ge r^2$$

and so

$$\lambda \ge \frac{r}{\sqrt{u^T H u}}.$$

Thus

$$\eta^T(x_0 - x') \ge \frac{\eta^T u}{\sqrt{u^T H u}}r$$
$$\ge \frac{\eta^T(x_0 - \omega)}{\sqrt{(x_0-\omega)^T H (x_0 - \omega)}}r \tag{5.1}$$

Note that

$$\psi_i(x_0) = \frac{a_i^T x_0 - a_i^T\omega}{(a_i^T\omega - b_i)}, \qquad i = 1, 2, \ldots, m,$$
$$\eta = -\sum_{i=1}^{m}\frac{1}{(a_i^T x_0 - b_i)}a_i,$$

and

$$H = \sum_{i=1}^{m}\frac{1}{(a_i^T x_0 - b_i)^2}a_i a_i^T.$$

Then

$$\eta^T(x_0 - \omega) = \sum_{i=1}^{m}\left(\frac{1}{1+\psi_i(x_0)} - 1\right)$$

and

$$(x_0 - \omega)^T H(x_0 - \omega) = \sum_{i=1}^{m} \left(\frac{1}{1 + \psi_i(x_0)} - 1 \right)^2.$$

So from (5.1) we get

$$\eta^T(x_0 - x') \geq \frac{\sum_{i=1}^{m} \left(\frac{1}{1 + \psi_i(x_0)} - 1 \right)}{\sqrt{\sum_{i=1}^{m} \left(\frac{1}{1 + \psi_i(x_0)} - 1 \right)^2}} r \tag{5.2}$$

Also,

$$f(x_0) = \sum_{i=1}^{m} \ln \left(\frac{1}{1 + \psi_i(x_0)} \right) \tag{5.3}$$

From Lemma 5.4,

$$\sum_{i=1}^{m} \psi_i(x_0) = 0.$$

So

$$\sum_{i=1}^{m} \left(\frac{1}{1 + \psi_i(x_0)} - 1 \right) = \sum_{i=1}^{m} \left(\frac{1}{1 + \psi_i(x_0)} + \psi_i(x_0) - 1 \right) = \sum_{i=1}^{m} \frac{\psi_i(x_0)^2}{1 + \psi_i(x_0)} \tag{5.4}$$

Also,

$$\sum_{i=1}^{m} \left(\frac{1}{1 + \psi_i(x_0)} - 1 \right)^2 = \sum_{i=1}^{m} \frac{\psi_i(x_0)^2}{(1 + \psi_i(x_0))^2}.$$

Thus from (5.2) we may conclude that

$$\eta^T(x_0 - x') \geq \frac{\sum_{i=1}^{m} \frac{\psi_i(x_0)^2}{1 + \psi_i(x_0)}}{\left(\sum_{i=1}^{m} \frac{\psi_i(x_0)^2}{(1 + \psi_i(x_0))^2} \right)^{1/2}} r \tag{5.5}$$

From (5.3) and (5.5) it follows that in order to prove the lemma it suffices to show that

(1) If $x_0 \notin \Sigma(\delta)$ then

$$\sum_{i=1}^{m} \frac{\psi_i(x_0)^2}{1 + \psi_i(x_0)} \geq \frac{\delta(1 - \delta)}{2(1 + \delta)} \left(\sum_{i=1}^{m} \frac{\psi_i(x_0)^2}{(1 + \psi_i(x_0))^2} \right)^{1/2}.$$

(2) If $x_0 \in \Sigma(\delta)$ then

$$\sum_{i=1}^{m} \frac{\psi_i(x_0)^2}{1 + \psi_i(x_0)} \geq \left((1 - \delta) \left(\sum_{i=1}^{m} \ln \left(\frac{1}{1 + \psi_i(x_0)} \right) \right) \left(\sum_{i=1}^{m} \frac{\psi_i(x_0)^2}{(1 + \psi_i(x_0))^2} \right) \right)^{1/2}.$$

Case 1. $x \notin \Sigma(\delta)$.

There are two subcases depending on the value of $\sum_{i=1}^{m} [\psi_i(x_0)^2/(1 + \psi_i(x_0))^2]$.

Case 1.1. $\displaystyle\sum_{i=1}^{m} \frac{\psi_i(x_0)^2}{(1+\psi_i(x_0))^2} \leq \delta^2.$

Since $x_0 \notin \Sigma(\delta)$, from (5.4) and Lemma 5.6 it follows that

$$\sum_{i=1}^{m} \frac{\psi_i(x_0)^2}{1+\psi_i(x_0)} \geq \frac{\delta^2}{1+\delta} \geq \frac{\delta}{1+\delta}\left(\sum_{i=1}^{m} \frac{\psi_i(x_0)^2}{(1+\psi_i(x_0))^2}\right)^{1/2} \tag{5.6}$$

Case 1.2. $\displaystyle\sum_{i=1}^{m} \frac{\psi_i(x_0)^2}{(1+\psi_i(x_0))^2} \geq \delta^2.$

Note that $1 + \psi_i(x) > 0$ for all points x in the interior of the polytope P. Thus

$$\sum_{|\psi_i(x_0)| \geq \delta} \frac{\psi_i(x_0)^2}{1+\psi_i(x_0)} \geq \sum_{|\psi_i(x_0)| \geq \delta} \delta \frac{|\psi_i(x_0)|}{|1+\psi_i(x_0)|} \geq \delta\left(\sum_{|\psi_i(x_0)| \geq \delta} \frac{\psi_i(x_0)^2}{(1+\psi_i(x_0))^2}\right)^{1/2}.$$

Suppose that

$$\sum_{|\psi_i(x_0)| \geq \delta} \frac{\psi_i(x_0)^2}{(1+\psi_i(x_0))^2} \geq \sum_{|\psi_i(x_0)| < \delta} \frac{\psi_i(x_0)^2}{(1+\psi_i(x_0))^2}.$$

Then it follows that

$$\sum_{|\psi_i(x_0)| \geq \delta} \frac{\psi_i(x_0)^2}{1+\psi_i(x_0)} \geq \frac{\delta}{\sqrt{2}}\left(\sum_{i=1}^{m} \frac{\psi_i(x_0)^2}{(1+\psi_i(x_0))^2}\right)^{1/2} \tag{5.7}$$

So let us assume that

$$\sum_{|\psi_i(x_0)| < \delta} \frac{\psi_i(x_0)^2}{(1+\psi_i(x_0))^2} \geq \sum_{|\psi_i(x_0)| \geq \delta} \frac{\psi_i(x_0)^2}{(1+\psi_i(x_0))^2}$$

Then

$$\sum_{|\psi_i(x_0)| < \delta} \frac{\psi_i(x_0)^2}{(1+\psi_i(x_0))^2} \geq \frac{\delta^2}{2}.$$

Thus

$$\sum_{|\psi_i(x_0)| < \delta} \frac{\psi_i(x_0)^2}{1+\psi_i(x_0)} \geq (1-\delta) \sum_{|\psi_i(x_0)| < \delta} \frac{\psi_i(x_0)^2}{(1+\psi_i(x_0))^2} \geq \frac{1}{\sqrt{2}}\delta(1-\delta)\left(\sum_{|\psi_i(x_0)| < \delta} \frac{\psi_i(x_0)^2}{(1+\psi_i(x_0))^2}\right)^{1/2} \geq \frac{1}{2}\delta(1-\delta)\left(\sum_{i=1}^{m} \frac{\psi_i(x_0)^2}{(1+\psi_i(x_0))^2}\right)^{1/2} \tag{5.8}$$

Thus from (5.6), (5.7), and (5.8) we may conclude that for Case 1

$$\sum_{i=1}^{m} \frac{\psi_i(x_0)^2}{1+\psi_i(x_0)} \geq \frac{\delta(1-\delta)}{2(1+\delta)} \left(\sum_{i=1}^{m} \frac{\psi_i(x_0)^2}{(1+\psi_i(x_0))^2} \right)^{1/2}.$$

Case 2. $x_0 \in \Sigma(\delta)$.

Note that since $x_0 \in \Sigma(\delta)$, $|\psi_i(x_0)| \leq \delta$, $i = 1, 2, \ldots, m$. So

$$\sum_{i=1}^{m} \frac{\psi_i(x_0)^2}{1+\psi_i(x_0)} \geq (1-\delta) \sum_{i=1}^{m} \frac{\psi_i(x_0)^2}{(1+\psi_i(x_0))^2}.$$

Also,

$$\begin{aligned}
\sum_{i=1}^{m} \ln\left(\frac{1}{1+\psi_i(x_0)}\right) &\leq \sum_{i=1}^{m} \left(\frac{1}{1+\psi_i(x_0)} - 1\right) \\
&\leq \sum_{i=1}^{m} \frac{\psi_i(x_0)^2}{1+\psi_i(x_0)} \qquad \text{(by (5.4))}
\end{aligned}$$

Thus for Case 2

$$\sum_{i=1}^{m} \frac{\psi_i(x_0)^2}{1+\psi_i(x_0)} \geq \left((1-\delta) \left(\sum_{i=1}^{m} \ln\left(\frac{1}{1+\psi_i(x_0)}\right) \right) \left(\sum_{i=1}^{m} \frac{\psi_i(x_0)^2}{(1+\psi_i(x_0))^2} \right) \right)^{1/2}$$

Lemma 5.3. *Let δ be a parameter such that $0 < \delta < 0.7$, and let ε be a parameter such that $0 < \varepsilon < 1$. Let r_0 be defined by*

$$r_0 = \begin{cases} \varepsilon, & \text{if } x_0 \notin \Sigma(\delta), \\ \varepsilon\sqrt{f(x_0)}, & \text{if } x_0 \in \Sigma(\delta). \end{cases}$$

Let $\hat{x}$ be the point that minimizes the linear function $\eta^T x$ over the ellipsoid $E(r_0)$ around x_0. The point $\hat{x}$ satisfies the following conditions.

(1) *If $x_0 \notin \Sigma(\delta)$ then*

$$f(\hat{x}) - f(x_0) \leq -\frac{\delta(1-\delta)}{2(1+\delta)}\varepsilon + 0.5\varepsilon^2 + \frac{\varepsilon^3}{3(1-\varepsilon)}.$$

(2) *If $x_0 \in \Sigma(\delta)$ then*

$$f(\hat{x}) \leq \left(1 - \varepsilon(1-\delta)^{1/2} + 0.5\varepsilon^2 + \frac{\varepsilon^3}{3(1-\varepsilon)}\right) f(x_0).$$

PROOF. Let x' be the point where the straight line joining x_0 and the center ω intersects the boundary of $E(r)$.

Case 1. $x_0 \notin \Sigma(\delta)$.

Proof by application of Lemmas 5.1 and 5.2 above and from the observation that $\eta^T \hat{x} \leq \eta^T x'$.

Case 2. $x_0 \in \Sigma(\delta)$.

Before we may apply Lemma 5.1 we must show that r_0 is less than 1. To show that $r_0 < 1$ it is adequate to prove that $f(x_0) \leq 1$. From Lemma 5.5, $f(x_0) \leq \delta^2/2(1 - \delta)$, and since $\delta \leq 0.7$ we get that $f(x_0) \leq 1$. We can now apply Lemmas 5.1 and 5.2, and noting that $\eta^T \hat{x} \leq \eta^T x'$ we may conclude that

$$f(\hat{x}) \leq f(x_0) - r_0\sqrt{(1-\delta)f(x_0)} + 0.5r_0^2 + \frac{r_0^3}{3(1-r_0)}$$

$$\leq \left(1 - \varepsilon\sqrt{(1-\delta)} + 0.5\varepsilon^2 + \frac{\varepsilon^3}{3(1-\varepsilon)}\right) f(x_0), \quad \text{as } f(x_0) < 1.$$

That concludes the proof for Case 2.

Acknowledgments

The author would like to thank Sanjiv Kapoor, Narendra Karmarkar, and Jeff Lagarias for helpful discussions and suggestions.

References

[1] D. A. Bayer, and J. C. Lagarias, The non-linear geometry of linear programming I. Affine and projective scaling trajectories, *Trans. Amer. Math. Soc.* (to appear).
[2] N. Karmarkar, A new polynomial time algorithm for linear programming, *Combinatorica* **4**(1984), 373–395.
[3] J. Renegar, A polynomial-time algorithm, based on Newton's method, for linear programming, MSRI 07118-86, Mathematical Sciences Research Institute, Berkeley, Calif., *Math. Programming* 1988, Volume 40, pages 59–94
[4] Gy. Sonnevand, An analytical center for polyhedrons and new classes of global algorithms for linear (smooth, convex) programming, Preprint, Department of Numerical Analysis, Institute of Mathematics, Eotvos University, Budapest.
[5] Gy. Sonnevand, A new method for solving a set of linear (convex) inequalities and its application for identification and optimization, Preprint, Department of Numerical Analysis, Institute of Mathematics, Eotvos University Budapest.
[6] P. M. Vaidya, An algorithm for linear programming which requires $O(((m+n)n^2 + (m+n)^{1.5}n)L)$ arithmetic operations, *Proc. ACM Annual Symposium or Theory of Computing* (1987), 29–38.

CHAPTER 6

A Note on Comparing Simplex and Interior Methods for Linear Programming

J. A. Tomlin

Abstract. This chapter discusses some aspects of the computational comparison of the simplex method and the new interior point (barrier) methods for linear programming. In particular, we consider classes of problems with which the simplex method has traditionally had difficulty and present some computational results.

§1. Introduction

Increased interest in linear programming (LP), stimulated by the publication of Karmarkar's [14] interior point method, continues at both a theoretical and computational level. There can be little doubt that the increased understanding of the relationship between the projective and affine scaling methods, Newton's method, and barrier methods (e.g., see Gill et al. [9]) has been beneficial in the development of new methods with a view to improved complexity bounds (e.g., Renegar [17]). Similarly, the investigation of "trajectories" of these methods (e.g., Megiddo [15]) has added refreshing geometric and intuitive insight into the LP solution process. What is not yet so clear is the actual computational benefit to be expected in practice from all these developments, even though much of the current interest was aroused by claims of orders of magnitude improvement over the simplex method.

Even in ideal circumstances, computational comparisons of different methods, or variants of the same method, can be difficult and the results difficult to reproduce. These problems were recognized some time ago, and guidelines were set down by the Mathematical Programming Society's Committee on Algorithms (COAL) and published (see Crowder et al. [5]). In 1985 a serious

This work was done while the author was at Ketron Management Science, Inc., Mountain View, CA 94040, USA.

attempt was made, under the auspices of the National Bureau of Standards and following these guidelines, to conduct independent comparisons of the simplex method and Karmarkar's method as implemented by AT&T Bell Laboratories. Unfortunately, the latter declined to cooperate (see Boggs et al. [4] and Garey and Gay [7]). In declining, Garey and Gay expressed the hope that the "evaluation ... that we all seek will occur naturally in the scientific community." To some extent this has come to pass, with computational results presented by (among others) Adler et al. [1, 2], Gill et al. [9], Goldfarb and Mehrotra [10]; Shanno and Marsten [18], and Tomlin [19]. However, none of these studies can yet be said to be definitive in the same way as the experiments of Hoffman et al. [12] were accepted as definitive in establishing the superiority of the simplex method over its (then) competitors.

It may be that none of the established mechanisms for computational comparison is adequate. The work involved in establishing practical "average case" behavior of the simplex versus various interior point (barrier) methods may simply be prohibitive, undefinable, or not in anyone's interest to carry out. On the other hand, ad hoc comparisons are vulnerable to charges that choice of a different algorithmic variant, data structure, programming language, or class of models could lead to (sometimes strikingly) different results.

The purpose of this chapter is therefore to sound a note of caution on the acceptance of computational results in this highly charged area and to present some such results, comment on others, and suggest a partial classification of problems that might be useful in computational evaluations.

§2. A Barrier Function Implementation

We intend to compare simplex and barrier methods in as unbiased a setting as possible and to this end are using implementations in a modern, state-of-the-art mathematical programming system, with all numerical routines coded for efficiency in assembly language. The details of the process of integrating a barrier method have been discussed by Tomlin and Welch [22] and will not be repeated here. The implementation uses the same data structures as the WHIZARD high-speed simplex code or a slight modification of the data structures described by George and Liu [8] where convenient. The only substantial difference from the implementation described in Tomlin and Welch [22] is that facilities have been added for dealing with bounds and ranges implicitly. That is, problems may be handled in standard MPS form:

$$\min_x \sum_j c_j x_j$$

subject to

$$Ax = b,$$

$$L_j \leq x_j \leq U_j.$$

The appropriate barrier function (see Gill et al. [9]) is then

$$\min_x \sum_j (c_j x_j - \mu \ln(x_j - L_j) - \mu \ln(U_j - x_j))$$

subject to

$$Ax = b$$

where $\mu > 0$ and $\mu \to 0$.

In outline, the algorithm for bounded models proceeds by defining

$$s_j = x_j - L_j, \qquad t_j = U_j - x_j,$$
$$D = \mathrm{diag}\{(1/s_j^2 + 1/t_j^2)^{-1/2}\},$$
$$\hat{D} = \mathrm{diag}\{(1/s_j - 1/t_j)\},$$
$$d = c - A^T\pi,$$
$$r = Dd - \mu D\hat{D}e.$$

Then the steps of an iteration are

(1) If μ and $\|r\|$ are sufficiently small—STOP.
(2) If "appropriate" reduce μ and recompute r.
(3) Solve a least squares problem:

$$\min_{\delta\pi} \|r - DA^T\delta\pi\|. \tag{6.1}$$

(4) Update the "pi values" and "reduced costs":

$$\pi \leftarrow \pi + \delta\pi, \qquad d \leftarrow d - A^T\delta\pi,$$

and compute the search direction p as

$$r = Dd - \mu D\hat{D}e, \qquad p = -(1/\mu)Dr.$$

(5) Calculate the step length α.
(6) Update $x \leftarrow x + \alpha p$. GO TO (1).

In practice, all finite lower bounds are translated by WHIZARD to zero. Free variables are treated by arbitrarily setting their bounds to be

$$x_j \le x_j^{(k)} + t\max\{|x_j^{(k)}|, 1\},$$
$$x_j \ge x_j^{(k)} - t\max\{|x_j^{(k)}|, 1\},$$

where $x^{(k)}$ is the current iterate. If t is chosen to be $\sqrt{2}$ it is easy to see that the diagonal entry in D turns out to be $\max\{|x_j^{(k)}|, 1\}$, while the entry in $\hat{D}$ is zero. The maximum step length calculation is of course modified to ignore free variables and take account of the upper bounds. We use an extra column length region to store D, now that it is no longer simply $\mathrm{diag}\{x_j\}$, but $\hat{D}$ is recomputed as needed.

The most critical step in each iteration remains the gradient projection, that is, solving the least squares problem (6.1). This involves the Cholesky factorization of an often approximate normal equation matrix

$$PLL^TP^T = \bar{A}\bar{D}^2\bar{A}^T \approx AD^2A^T, \tag{6.2}$$

where the permutation matrix P is chosen so that the lower triangular factor L is sparse. This is used to solve the preconditioned problem

$$\min_y \|r - DA^TPL^{-T}y\| \tag{6.3}$$

and recover $\delta\pi = PL^{-T}y$. It is the operations involving L that use the data structures from George and Liu [8]. All operations involving the original problem data use the WHIZARD data structure. The algorithm control sequence, tolerances, etc. follow closely those given in Gill et al. [9].

§3. Simplex Implementation

Aspects of the WHIZARD simplex implementation have been described in several places in the literature. It uses a super-sparse storage scheme (Kalan [13]), the Hellerman-Rarick [11] P^4 inversion, and Forrest-Tomlin [6] updating. Pure network problems are automatically detected and dealt with using the WHIZNET primal network simplex implementation (see Tomlin and Welch [21]). Structurally degenerate rows and columns, that is, those identifying "null variables" via constraints of the form

$$\sum_j a_j x_j \leq 0,$$

$$\text{all } a_j \geq 0,\ x_j \geq 0,$$

and "implied free variables" y identified by constraints of the form

$$\sum_j a_j x_j - ay \leq b,$$

$$\text{all } a_j \geq 0,\ x_j \geq 0,\ y \geq 0,\ b \leq 0,$$

are detected and dealt with by a PRESOLVE routine, before "CRASHing" a starting basis. Formal optimality for the original model is enforced by POSTSOLVE (see Tomlin and Welch [20]).

The WHIZARD pricing strategy is to segment the matrix into 16 (fixed) equal parts and attempt to find 5 promising candidates for the basis, no more than one per segment. When a satisfactory candidate is found in a segment the remainder is ignored and the next segment to be examined is chosen in a crude pseudorandom way. The idea, of course, is to avoid choosing closely correlated columns for the suboptimization.

Most of the comparisons between the simplex and interior methods given so far have used MINOS (Murtagh and Saunders [16]) as a simplex benchmark, or occasionally the IBM MPSX/370 program product (see Benichou et al. [3]). While both of these are widely distributed, neither represents a current state-of-the-art simplex implementation. MINOS is a Fortran code intended primarily for the numerically stable solution of nonlinear problems (although it does employ many advanced sparse matrix techniques), whereas MPSX is of a more traditional out-of-core design. MINOS has no antidegeneracy or PRESOLVE mechanism. Both are slower than WHIZARD on most problems.

§4. Initial Experiments

The test problems to be discussed are described in Table 6.1. The "reduced" row and column numbers reflect fixed and null variables and redundant rows determined by PRESOLVE. Except where noted, the other statistics are for the reduced problems. The other statistics are as follows: F gives the number of original plus implied free variables, U is the number of bounded variables (including ranged logicals), A is the number of original problem nonzeros (including logicals, right-hand-side nonzeros, and the cost row), B is the number of off-diagonal nonzeros in the LU factors of the optimal basis, S is the number of spikes in that basis, Adj denotes the number of nonzeros below the diagonal in AA^T, and $|L|$ denotes the number of nonzeros below the diagonal of L.

Our initial experiments were with the first nine problems, often referred to as the "Stanford test set." Note that most of the larger models have some redundancy detected by PRESOLVE. The ISRAEL problem is interesting because it has six dense columns removed when forming the preconditioner via (6.2). This results in at least seven conjugate gradient iterations during the projections. The results of these runs are given in Table 6.2. In all cases the times given include model input, conversion to WHIZARD format, PRESOLVE, and all other overhead. The simplex method times include POSTSOLVE time (if any). The barrier method times include symbolic factorization and minimum degree ordering time, which is never more than about 8% of the total for this first sample. They do *not*, however, include any time to produce a final basic solution. The importance of this step was emphasized in Tomlin and Welch [22] and it may require substantial work, but it is not considered here.

It is clear that the barrier method is inferior to the simplex method for this sample and would have to improve considerably to outperform it. It might be retorted that these problems are too small, or too easy, to be a fair test. Though not large, these problems are of realistic structure and seem reasonably representative of many small- to medium-scale applications. Given this, we might ask what kind of experiments would make the barrier method look attractive.

There are at least three ways to accomplish this. One would be to use an unsuitable implementation of the simplex method. It has already been noted that some other experimenters are using slower simplex codes. Another op-

Table 6.1. Test Problems

Problem name	Original rows	Original columns	Reduced rows	Reduced columns	F	U	A	B	S	Adj	$\lvert L \rvert$
AFIRO	28	32	28	32	16	0	116	40	0	63	80
SHARE2B	97	79	97	79	6	0	901	570	20	775	930
SHARE1B	118	225	118	225	6	0	1300	673	28	855	1234
BEACONFD	174	262	124	184	20	0	3650	719	0	1603	1610
ISRAEL	175	142	175	141	0	0	2533	1475	3	3437	3552
BRANDY	221	249	131	222	11	0	2371	1199	16	1944	2552
E226	224	282	199	262	20	0	3264	1323	13	2255	2984
CAPRI	272	353	272	337	22	131	2058	1412	19	2621	4770
BANDM	306	472	244	322	33	0	2965	1762	27	2658	3843
NZFRI	624	3521	415	2034	0	0	16527	1498	8	4172	9055
MTZ	1146	1216	554	445	194	0	6589	1677	21	2686	3853
SCSD6	148	1350	148	1350	0	0	5814	677	6	1952	2398
SCSD8	398	2750	398	2750	0	0	11732	1644	29	3883	5482
DEGEN1	67	72	67	72	21	0	411	229	1	326	500
DEGEN2	445	534	442	534	3	0	4894	3709	32	6838	15863
DEGEN3	1504	1818	1504	1818	10	0	27734	20687	101	50178	119300
ORGDES	125	595	125	595	0	0	1910	368	0	595	605
NETGEN17	401	2443	401	2443	2	0	7730	—	0	2425	26394
NETGEN27	401	2676	401	2676	0	2111	8429	—	0	2652	29619
SHELL	537	1775	537	1525	194	117	5437	1488	0	1406	3854
APL	141	127	141	127	127	140	2438	2223	125	1841	1841
STAIR	357	467	357	385	70	6	4214	12097	144	6197	16685
BP822	822	1571	822	1571	113	0	11127	8627	132	10758	32880

Table 6.2. Stanford Text Problems

	WHIZARD PRIMAL		WHIZARD BARRIER	
Problem name	Iterations	Time*	Iterations	Time
AFIRO	11	0.06	20	0.18
SHARE2B	110	0.36	22	0.96
SHARE1B	152	0.72	40	2.22
BEACONFD	22	0.42	28	3.72
ISRAEL	199	1.38	43	13.2
BRANDY	168	1.38	27	3.72
E226	321	1.80	34	5.22
CAPRI	270	1.02	33	6.12
BANDM	253	1.86	32	4.92

* All times are in seconds on an Amdahl 5860.

tion would be to choose problems that are highly favorable to barrier-type methods. These would include models where the (appropriately ordered) matrix AA^T has a structure particularly amenable to Cholesky factorization. Such models would arise, for example, in multicommodity network models where the number of commodities is large in comparison to the network size (it seems to have been such models that AT&T used to benchmark their interior point code against MPSX). Third, one might look for problems that the simplex method finds difficult. This seems much the most constructive approach, if we are genuinely interested in increasing problem-solving power.

§5. Difficulty of Linear Programs

Several factors have been cited as making some LP models hard (expensive) to solve:

(a) A large number of rows. In practice, the old rule of thumb of $2m$ iterations for an m-row model is quite good for most models. Because of sparsity, the actual work tends to increase at nothing like the theoretical $O(m^3)$ rate, but rather $O(m^p)$ with p more nearly in the range $1.5 \leq p \leq 2$. Thus, unless the model displays some other difficult characteristic, mere size is not necessarily a problem.

(b) A large number of variables. While complexity theorists are accustomed to using functions of n as a measure of complexity, this is usually significant in LP only if n is very much larger than m. Such models usually display some special kind of structure (e.g., GUB or embedded network) and are often highly degenerate.

(c) Degeneracy. While actual cycling is almost unheard of in practice, almost all models exhibit some primal and/or dual degeneracy. This can result in an enormous number of simplex iterations, almost all of which lead to no change in the objective. Thus the difficulty with many models is not that the simplex method visits so many vertices, as is so often propounded, but that it visits rather few and dwells at them overlong. Serious degeneracy seems to arise in two flavors:

(i) Structural degeneracy of the type that can be dealt with by PRESOLVE/POSTSOLVE.

(ii) "Hard" degeneracy inherent in the model.

(d) Unusually high density. Most practical LP models have an average of four to six nonzeros per column. Formulations with higher density—even as few as seven or eight nonzeros per column—seem to be much more difficult, requiring an unexpectedly high number of iterations and significantly increased work per iteration.

(e) Special structure. Many special structures can be exploited to considerable benefit, but some structures are more difficult to deal with. In particular, multistage (staircase) problems usually require an inordinate number of iterations to solve—sometimes up to $10m$.

(f) Numerical instability. It is often difficult to reconcile speed and a priori numerical stability; therefore compromises are made that may require relatively expensive a posteriori correction when breakdown does occur.

All of these phenomena play some part in the experiments we shall discuss, but degeneracy and special structure are perhaps the most important.

§6. Degenerate Models

Our test problems include a set of structurally degenerate models (NZFRI and MTZ) and a set of "intrinsically" degenerate models (the SCSD and DEGEN models). Computational results for the first group are given in Table 6.3 for both the reduced and unreduced models. In this table we give the B,

Table 6.3. Structurally Degenerate Models

	WHIZARD PRIMAL		WHIZARD BARRIER					
Problem name	Iterations	Time	Iterations	Time	B	S	Adj	$\|L\|$
NZFRI (Unreduced)	5496	33.84	43	38.40	2832	8	7164	17974
NZFRI (Reduced)	576	6.00	34	18.24	1498	8	4172	9055
MTZ (Unreduced)	1567	9.84	36	18.42	3972	21	6567	11355
MTZ (Reduced)	139	1.38	32	6.24	1677	21	2686	3853

S, Adj, and $|L|$ statistics for both forms of each model, since Table 6.1 gives these figures only for the reduced forms. We see that removal of the structural degeneracy (null variables) has a nearly tenfold effect on the number of simplex iterations for NZFRI and even more for MTZ. The difference in times is fivefold and more. On the other hand, the barrier method reaps only a small advantage in number of iterations and a two- or threefold advantage in times from the same reductions. We also see that the barrier method on the reduced problems is faster than the simplex method on the unreduced problems, while direct comparison for both cases is advantageous to the simplex method.

It should also be pointed out that NZFRI is a forest planning model and that Adler et al. [2] have noticed another anomaly with such models. Comparing their Fortran dual affine scaling code with MINOS 4.0, they found that MINOS could go from looking markedly inferior if full pricing was used to consistently better if "optimal" partial pricing parameters were used. It is thus clear that comparisons for this type of problem must be very carefully made to be meaningful.

The SCSD problems are highly degenerate models derived from structural analysis. They also have a staircase structure. PRESOLVE is unable to detect any null variables (in either these or the DEGEN problems). For this class of models the barrier method does lead to an improvement that seems to grow with problem size—perhaps quite rapidly (see Table 6.4). One factor not evident from the table is that numerical difficulties were encountered by the WHIZARD fast primal algorithm with SCSD8, requiring recourse to the slower but more stable primal variant.

The DEGEN problems are also massively degenerate and quite dense—another bad omen. We see that the barrier method has produced a marked improvement for DEGEN2. At first sight it is disappointing that the barrier performance for DEGEN3 is worse than the simplex method. However, this failure to continue an improving trend seems to be due not to degradation of the barrier performance with model size but to the fact that the simplex method managed to solve the problem in comparatively fewer iterations. Many larger models of this type display a much worse iteration-to-row-size ratio than this model. Given the fact that we can still make some speed

Table 6.4. "Hard" Degenerate Models

	WHIZARD PRIMAL		WHIZARD BARRIER	
Problem name	Iterations	Time	Iterations	Time
SCSD6	924	4.44	20	3.78
SCSD8	3374	37.8	20	7.32
DEGEN1	22	0.12	16	0.36
DEGEN2	3995	35.58	24	18.54
DEGEN3	12033	285.1	34	330.8

Table 6.5. Network Models

	WHIZARD PRIMAL		WHIZARD BARRIER	
Problem name	Iterations	Time	Iterations	Time
ORGDES	79	0.42	32	1.62
NETGEN17	1218	0.84	30	57.96
NETGEN27	1774	1.02	33	78.12
SHELL	331	1.92	40	8.76

improvements to the barrier code, we must count this class of models as a success for it.

§7. Network Models

Another class of models that is highly degenerate and displays special structure is that of network problems. Obviously the normal matrix in (6.2) will have special form for pure (and generalized) networks. Various observers have suggested that, even without taking advantage of this, an interior point method would outperform the network primal simplex method. Our preliminary results (see Table 6.5) do not confirm this. Problem ORGDES is a transportation problem with only 5 sources and 119 destinations. Although we have used the regular primal algorithm (since it was not set up in transshipment form), it outperforms the barrier method. Problems NETGEN17 and NETGEN27 were produced by the well-known network model generator and are transshipment problems. The WHIZNET code (see [21]) far outperforms our barrier implementation, which, incidentally, is able to take advantage of the super-sparsity of the data. Note the massive fill-in of the Cholesky factors for both models.

Problem SHELL is a network problem with joint capacity constraints and thus unable to use WHIZNET directly. Despite this, the simplex algorithm proved faster in this case. This is somewhat surprising, since multicommodity network models (with joint capacities) have been one of the success stories for the AT&T implementation reported in the press [23], as we mentioned above. Other results for such problems are presented by Adler et al. [2], who found that their Karmarkar code was much faster than MINOS on multicommodity flow problems but often slower than a specialized simplex code by Kennington. This then begs the question of how much an interior point method gains from specialization.

§8. Other Structures

One structure on which the simplex method tends to perform badly is the band matrix, usually arising from finite-difference problems in physics and engineering. Most simplex codes will try to exploit triangularity, pivoting on

Table 6.6. Miscellaneous Structures

	WHIZARD PRIMAL		WHIZARD BARRIER	
Problem name	Iterations	Time	Iterations	Time
APL	277	2.76	21	2.04
STAIR	344	12.60	45	40.26
BP 822	2860	86.22	47	83.76

off-diagonal elements, with unfortunate results. The barrier method, on the other hand, finds the structure ideal. Problem APL is an electrical engineering model with band structure. All the structural variables are free, and the rows are ranged. Note that there is no fill-in of the Cholesky factors, and the barrier method is able to outperform the simplex method even for this relatively small size (see Table 6.6). We would expect the advantage to grow with problem size.

Most multistage models, as we commented in Section 5, require a great many simplex iterations to solve. Problem STAIR is a multistage model but solves in only $\sim m$ iterations. The barrier method, however, took a fairly typical number of iterations. If this model had needed anything like the usual large number of simplex iterations (say $5m$ to $10m$), we see that the barrier method would indeed be faster, and thus we should probably regard the actual result obtained here as unrepresentative.

Finally, we consider the notorious BP 822 row model (sometimes called, less evocatively, 25FV47). This has been a "code-breaker" test problem since the 1960s. Many simplex codes have taken many thousands of iterations to solve it. It is dual-angular and dense, with about seven nonzeros per column. The results we obtained with the WHIZARD simplex method were therefore quite pleasing. It was then most impressive to find that our barrier code had succeeded in reaching optimality in slightly better time—one of the most encouraging signs yet that this method would indeed be of real help in dealing with difficult problems.

§9. Conclusion

In assessing the value of new LP techniques, such as interior point methods, there are many issues to be considered. Among them are stability, ease of use, restart efficiency, compatibility with other mathematical programming procedures, and usability of the solutions, as well as raw speed. It must be stressed that this chapter and the computational results are relevant only to the latter.

While the actual computational results achieved to date have been mixed, there are definite indications that interior (barrier) methods can be expected to add substantively to problem-solving power. Perhaps even better, it appears that this new power complements rather than duplicates or supersedes the efficiency of the simplex method on the problems it handles well, being most applicable to models with which the simplex method has difficulty. It is suggested that closer identification of problem classes and algorithm suitabilities might be a more productive avenue of investigation than very general claims whose meaning is difficult to evaluate.

Acknowledgments

I am indebted to Philip Gill, Walter Murray, and Christopher Strauss for helpful discussions in the course of this work and to an anonymous referee for constructive criticism.

References

[1] I. Adler, M. G. C. Resende, and G. Veiga, An implementation of Karmarkar's algorithm for linear programming, Report ORC 86-8, Department of IE/OR, University of California, Berkeley (1986).

[2] I. Adler, N. Karmarkar, M. G. C. Resende, and G. Veiga, An implementation of Karmarkar's algorithm for linear programming, Presentation at the 22nd TIMS/ORSA meeting, Miami (1986).

[3] M. Benichou, J. M. Gauthier, G. Hentges, and G. Ribière, The efficient solution of large-scale linear programming problems—some algorithmic techniques and computational results, *Math. Programming* **13** (1977), 280–322.

[4] P. Boggs, K. Hoffman, and R. Jackson, NBS group report to MPS chairman, *COAL Newslett.*, No. 13 (1985), 5–7.

[5] H. P. Crowder, R. S. Dembo, and J. M. Mulvey, On reporting computational experiments with mathematical software, *ACM Trans. Math. Software* **5** (1979), 193–203.

[6] J. J. H. Forrest and J. A. Tomlin, Updating triangular factors of the basis to maintain sparsity in the product form simplex method, *Math. Programming* **2** (1972), 263–278.

[7] M. R. Garey and D. M. Gay, Letter from AT&T Bell Laboratories, *COAL Newslett.*, No. 13 (1985), 3.

[8] J. A. George and J. W. Liu, *Computer Solution of Large Sparse Positive Definite Systems*, Prentice-Hall, Englewood Cliffs, N.J., 1981.

[9] P. E. Gill, W. Murray, M. A. Saunders, J. A. Tomlin, and M. H. Wright, On projected Newton barrier methods for linear programming and an equivalence to Karmarkar's projective method, *Math. Programming* **36** (1986), 183–209.

[10] D. Goldfarb and S. Mehrotra, A relaxed version of Karmarkar's method, *Math. Programming* **40** (1988), 289–315.

[11] E. Hellerman and D. C. Rarick, Reinversion with the partitioned preassigned pivot procedure, in *Sparse Matrices and Their Applications*, D. J. Rose and R. A. Willoughby (eds.), Plenum Press, New York, 1972, pp. 67–76.

[12] A. J. Hoffman, M. Mannos, D. Sokolowsky, and N. Wiegmann, Computational experience in solving linear programs, *SIAM J.* **1** (1953), 1–33.

[13] J. E. Kalan, Aspects of large-scale in-core linear programming, *Proc. ACM Annual Conference*, Chicago, ACM, New York, 1971, pp. 304–313.

[14] N. Karmarkar, A new polynomial-time algorithm for linear programming, *Combinatorica* **4** (1984), 373–395.

[15] N. Megiddo, Pathways to the optimal set in linear programming. In: *Progress in Mathematical Programming: Interior-Point and Related Methods*. Springer-Verlag, New York (1989).
[16] B. A. Murtagh and M. A. Saunders, MINOS 5.0 User's Guide, Report SOL 83-20, Department of Operations Research, Stanford University (1983).
[17] J. Renegar, A polynomial-time algorithm, based on Newton's method, for linear programming, *Math. Programming* **40** (1988), 59–93.
[18] D. F. Shanno and R. E. Marsten, On implementing Karmarkar's method, Working Paper 85-01, Graduate School of Administration, University of California, Davis (1985).
[19] J. A. Tomlin, An experimental approach to Karmarkar's projective method for linear programming, *Math. Programming Studies* **31** (1987), 175–191.
[20] J. A. Tomlin and J. S. Welch, Formal optimization of some reduced linear programming problems, *Math. Programming* **27** (1983), 232–240.
[21] J. A. Tomlin and J. S. Welch, Integration of a primal simplex network algorithm with a large-scale mathematical programming system, *ACM Trans. Math. Software* **11** (1985), 1–11.
[22] J. A. Tomlin and J. S. Welch, Implementing an interior point method in a mathematical programming system, presented at the 22nd TIMS/ORSA meeting, Miami (1986).
[23] *Wall Street Journal*, Karmarkar algorithm proves its worth, July 18, 1986.

CHAPTER 7

Pricing Criteria in Linear Programming

J. L. Nazareth

Abstract. In this chapter we discuss gradient-based descent methods for solving a linear program. The definition of a local reduced model and selection of a suitable direction of descent provide the overall framework within which we discuss some existing techniques for linear programming and explore other new ones. In particular, we propose a null space affine (scaling) technique that is motivated by the approach of Karmarkar (1984) but builds more directly on the simplex method. We discuss algorithmic considerations based on this approach and issues of effective implementation, and we report the results of a simple, yet instructive, numerical experiment.

§1. Introduction

1.1. Overview

In this chapter we discuss gradient-based descent methods for solving a linear program. At each iteration, such methods form a reduced model of the linear program and use this model to find a suitable direction of descent, thereby obtaining an improved approximation to the solution.

Definition of a *local reduced model* and selection of a suitable *direction of descent* are operations that are generically known by the name "pricing" in the context of the simplex method (hence our chapter's title). They provide the overall framework within which we discuss some existing techniques for linear programming and explore other new ones (Section 2). Our main contribution in this section is a null space affine (scaling) technique that is motivated

Research supported, in part, under ONR Contract No. N00014-85-K-0180.

by the approach of Karmarkar [28] but builds more directly on the simplex method (see Section 2.4).

Algorithmic considerations based on this approach and issues of effective implementation are the subject of Section 3. In particular, we consider how a key system of linear equations that arise at each iteration can be solved to the necessary accuracy, and we discuss the conditioning of this system [see (7.35)].

In Section 4 we report the results of a numerical experiment that is designed to investigate the viability of our approach on a set of small dense (Kuhn-Quandt) linear programming problems.

In Section 5 we briefly discuss how simplex-like and Karmarkar-like local reduced models can be combined.

Practical considerations and connections with other affine (scaling) algorithms form the subject of Section 6.

Finally, in Section 7, some observations in a philosophical vein conclude our chapter.

1.2. Context

Khachiyan's [30] theoretical breakthrough, which was based on the convex programming techniques of Shor [51] and others, demonstrated the value of adapting iterative methods of nonlinear programming and nonlinear equation solving to the special case of solving a linear program. The polynomial-time algorithm of Karmarkar [28] which was presented from the novel standpoint of projective geometry, was soon realized also to have close ties to nonlinear programming (NLP) techniques; see remarks in Todd and Burrell [56] and the detailed study of Gill et al. [19]. (Earlier, interesting but less dramatic attempts to apply NLP techniques to linear programming were proposed, for example, by Golshtein [22] and Mangasarian [35]. More recent approaches based on nonlinear equation solving are given, for example, by Smale [53] and Blum [8].) The net result of these and other contributions has been to steer linear programming away from combinatorial programming and to place the subject more squarely within the realm of nonlinear programming that simultaneously seeks to take advantage of the special characteristics of a linear program.

Algorithmic techniques for nonlinear programming, particularly when constraints are linear, fall into two broad categories, namely techniques that produce successive iterates with monotonic objective function values and techniques that produce successive iterates with nonmonotonic values. Null space projective (scaling) versions of Karmarkar's method are given by Shanno [50] and by Goldfarb and Merothra [20] and generate non-monotonic iterates. Anstreicher [2] gives a monotonic version of Karmarkar's algorithm. The affine (scaling) algorithm is a gradient-based *inherently* monotonic variant of Karmarkar's projective (scaling) algorithm that was proposed by numerous researchers after Karmarkar's paper appeared, among them

Vanderbei et al. [59], Barnes [4], and Chandru and Kochar [9]; see also Nazareth [40] (homotopy approach), Ye [61] (region of trust approach), and Strang [55].

Computational viability of the approach and its convergence under nondegeneracy assumptions were demonstrated by Vanderbei et al. [59]. The null space affine (scaling) variant is presented here, our particular motivation being our earlier work described in Nazareth [40]. We also present the encouraging results of numerical experiments on a small set of Kuhn-Quandt *dense* LP problems (see Avis and Chvatal [3]), which were obtained with an implementation that computes descent search directions *inexactly* by the conjugate gradient method. Algorithm I of Adler et al. [1], derived in a different manner, can also be shown to be a particular null space affine (scaling) algorithm. They report encouraging computational results on a set of *large sparse* problems, using an implementation that (usually) computes search directions *exactly* by sparse Cholesky factorization. See also Tone [57] for related work. Megiddo and Shub [34] show that the affine (scaling) algorithm can "hug the boundary" and exhibit Klee-Minty type behavior, and Barnes et al. [5] show how to interleave steps of the affine (scaling) algorithm with "centering" steps to give an algorithm that restores the polynomial bound. Whether there is a *gradient-based inherently monotonic* polynomial-time algorithm for linear programming remains an open question.

This chapter deals with gradient-based monotonic algorithms and the use of NLP techniques in linear programming. The simplex method is gradient-based monotonic, and the null space affine (scaling) variant of Karmarkar's method introduced here meshes well with it. An approach that combines simplex-like and Karmarkar-like local reduced models is briefly outlined in Section 5.

§2. A Progression of Methods

2.1. Interpretation of the Simplex Method

Consider a linear program in standard form

$$\begin{aligned} &\text{minimize } c^T x \\ &\text{s.t. } Ax = b, \\ &x \geq 0, \end{aligned} \tag{7.1}$$

where A is an $m \times n$ matrix of full rank and $m \leq n$. Let x^0 be a feasible solution of (7.1) and define $z^0 \equiv c^T x^0$ and $\bar{n} \equiv n - m$. We shall assume that the feasible region of (7.1) is bounded with a nonempty interior.

Let Z be an $n \times (n - m)$ matrix of full rank for which

$$AZ = 0, \tag{7.2}$$

and make a transformation of variables

$$x = x^0 + Zw. \tag{7.3}$$

Then (7.1) is equivalent to the *reduced problem*:

$$\begin{aligned} &\text{minimize } (Z^T c)^T w + z^0 \\ &\text{s.t. } x^0 + Zw \geq 0. \end{aligned} \tag{7.4}$$

In the *simplex method*, Z is defined to be

$$Z \equiv \begin{bmatrix} -B^{-1}N \\ I_{\bar{n}\times\bar{n}} \end{bmatrix}, \tag{7.5}$$

where $A = [B \mid N]$ is partitioned in the usual manner into an $m \times m$ nonsingular basis matrix B (without loss of generality it can be taken to be the first m columns of A) and an $m \times \bar{n}$ matrix N of nonbasic columns. $I_{\bar{n}\times\bar{n}}$ denotes the $\bar{n} \times \bar{n}$ identity matrix. Clearly (7.5) satisfies relation (7.2). Correspondingly, partition the variables $x = \begin{bmatrix} x_B \\ x_N \end{bmatrix}$ and the feasible solution $x^0 = \begin{bmatrix} x_B^0 \\ x_N^0 \end{bmatrix}$ and define $\Delta x_N \equiv x_N - x_N^0$. From (7.3), $x_B = x_B^0 - B^{-1}Nw$ and $\Delta x_N = w$. Let us also define $\sigma_N^0 \equiv Z^T c$. (This quantity is often called the *reduced gradient* for the following reason: Given any differentiable objective function $f(x)$, a straightforward application of the chain rule shows that its gradient $\nabla f(x)$ transforms to $Z^T \nabla f(x)$ under the change of variables (7.3). In our case $f(x) = c^T x$ and $\nabla f(x) = c$.) Thus we can express (7.4) as follows.

$$\begin{aligned} &\text{minimize } (\sigma_N^0)^T \Delta x_N + z^0 \\ &\text{s.t. } x_B^0 - B^{-1}N\Delta x_N \geq 0,\ x_N^0 + \Delta x_N \geq 0. \end{aligned} \tag{7.6}$$

Assume x_B^0 to be nondegenerate, that is, $x_B^0 > 0$. Then, for sufficiently small values of Δx_N, we can also ensure that $x_B^0 - B^{-1}N\Delta x_N > 0$. *Locally therefore*, that is, in the neighborhood of x^0, the first set of constraints in (7.6) can be dropped, leading to the *local reduced model*:

$$\begin{aligned} &\text{minimize } (\sigma_N^0)^T \Delta x_N + z^0 \\ &\text{s.t. } x_N^0 + \Delta x_N \geq 0. \end{aligned} \tag{7.7}$$

(Note that a more elementary derivation of (7.7) is to directly substitute the relations $x_B = B^{-1}b - B^{-1}Nx_N$ and $x_B^0 = B^{-1}b - B^{-1}Nx_N^0$ into (7.1). Our choice of the above line of development is geared to the needs of subsequent discussion.)

In the simplex method a *single coordinate* in the reduced space, say $e_k(\Delta x_N)_k$ (where e_k denotes a unit vector in R^{n-m}) is chosen so that the corresponding component of the reduced gradient σ_N^0 satisfies

$$(\sigma_N^0)_k < 0 \quad \text{when } (x_N^0)_k = 0 \qquad \text{or} \qquad (\sigma_N^0)_k \neq 0 \quad \text{when } (x_N^0)_k > 0. \tag{7.8}$$

The index k is usually selected to be the one for which $(\sigma_N^0)_k$ is largest in absolute value subject to (7.8). The sign of $(\Delta x_N)_k$ is chosen to ensure function decrease, that is,

$$(\sigma_N^0)_k(\Delta x_N)_k < 0. \tag{7.9}$$

The associated direction in the original space is defined by

$$\Delta x = x - x^0 = Ze_k(\Delta x_N)_k = z_k(\Delta x_N)_k, \tag{7.10}$$

where z_k is the kth column of Z. Expression (7.9) ensures that Δx is a direction of descent because $c^T\Delta x = c^T z_k(\Delta x_N)_k = (\sigma_N^0)_k(\Delta x_N)_k < 0$. The corresponding step along Δx is then chosen so as to obtain as large an improvement as possible, the basis is revised, and the cycle is repeated.

We see that the simplex method can be interpreted as a *method of coordinate descent* in a reduced space (which is revised at each iteration). *Note that we have made no assumption that nonbasic variables are at a bound and when* x^0 *is an interior point, the path generated will lie in the interior and on the facets of the feasible polytope.* This extension of the usual simplex procedure is useful in practice. Related considerations are given by Zangwill [62] and Solow [54] and turn out to be a specialization of the superbasic variables of Murtagh and Saunders [38]. For a further discussion see Nazareth [44].

2.2. Steepest (Reduced) Coordinate Pricing and Related Extensions

The matrix Z of (7.2) and (7.5) is by no means unique. Let us require that each column of Z in (7.5) be of unit (Euclidean) length by defining

$$\bar{Z} \equiv Z\,\mathrm{diag}[1/\|z_1\|_2, \ldots, 1/\|z_{n-m}\|_2], \tag{7.11}$$

where $\|\cdot\|_2$ denotes the Euclidean norm. From (7.2), $A\bar{Z} = 0$. The corresponding reduced gradient $\bar{\sigma}_N^0 \equiv \bar{Z}^T c$ has components

$$(\bar{\sigma}_N^0)_k = c^T z_k/\|z_k\|_2. \tag{7.12}$$

This is the *directional derivative* along z_k and a choice of descent direction based on $\bar{\sigma}_N^0$ (instead of σ_N^0) defines *steepest (reduced) coordinate pricing*. When x^0 is a basic feasible solution, this corresponds to *steepest-edge pricing* of Goldfarb and Reid [21]. Greenberg and Kalan [25] independently give a slight variant where (7.11) is replaced by

$$\hat{Z} \equiv Z\,\mathrm{diag}[1/\|B^{-1}n_1\|_2, \ldots, 1/\|B^{-1}n_{n-m}\|_2], \tag{7.13}$$

$$(\hat{\sigma}_N^0)_k \equiv c^T z_k/\|B^{-1}n_k\|_2, \tag{7.14}$$

where n_k is the kth column of N. Note that $\|z_k\|_2 = (\|B^{-1}n_k\|_2^2 + 1)^{1/2}$ and again $A\hat{Z} = 0$. Expression (7.14) has the advantage that it is invariant with respect to row scaling and with respect to scaling of the *nonbasic* variables. Greenberg and Kalan [25] and Goldfarb and Reid [21] also independently proposed an effective technique for updating the quantities $\|B^{-1}n_k\|_2$, or equivalently $\|z_k\|_2$, from one iteration of the simplex method to the next. An elementary derivation of this technique may be found in Nazareth [44]. Approximations to these quantities, which can be computed much more efficiently, are given by Harris [26] and by Benichou et al. [7].

The *norm* in (7.11) is still open to alternative choice. One appealing possibility, when x_B^0 is nondegenerate, is to utilize the norm $\|\cdot\|_{D_0^{-2}}$ defined by

$$\|v\|_{D_0^{-2}} \equiv (v^T D_0^{-2} v)^{1/2} \tag{7.15}$$

where $D_0 \equiv \text{diag}[x_1^0, \ldots, x_m^0] > 0$, and to define quantities analogous to (7.14) as follows.

$$(\tilde{\sigma}_N^0)_k \equiv c^T z_k / \|B^{-1} n_k\|_{D_0^{-2}}. \tag{7.16}$$

It can then easily be verified that (7.16) is invariant with respect to row scaling and with respect to (column) scaling of *both* basic variables and nonbasic variables. It is also easily verified that (7.16) could equivalently be derived by a Karmarkar-like transformation of variables involving only the *basic* variables,

$$x_B = D_0 \tilde{x}_B, \qquad x_N = \tilde{x}_N, \tag{7.17}$$

and definition of quantities (7.14) on the transformed linear program. Preliminary numerical experimentation has shown us that choices based on (7.12) or (7.14) lead to better performance of the simplex algorithm. These latter quantities are also more efficient to compute and the practical utility of (7.16) is therefore unclear. A further discussion of scale-invariant criteria may be found in Greenberg [24] and Nazareth [44].

2.3. Projected and Reduced Gradient Extensions

We have noted that the simplex method utilizes coordinate directions in a reduced space. It would thus be natural to consider changes in several coordinates simultaneously, in particular when x^0 is an interior point. One convenient approach to defining a local reduced model that is well suited to this purpose is through the use of a quadratic perturbation or *quadratic regularizing term* in the objective function of (7.1). The device of a quadratic regularizing term, usually in a Euclidean metric, was pioneered by Pschenichny (see Pschenichny and Danilin [46]), and it is now widely used in convex programming, see Rockafellar [48]. Mangasarian [35] employs a quadratic perturbation of a linear program, which is then solved in the space of dual variables by SOR techniques. See Nazareth [40] for another example of its use in linear programming.

Given a feasible point x^0 with $z^0 \equiv c^T x^0$, we can reexpress the linear program (7.1) in the following form:

$$\begin{aligned} &\text{minimize } c^T(x - x^0) + z^0 \\ &\text{s.t. } A(x - x^0) = 0, \ x \geq 0. \end{aligned} \tag{7.18a}$$

Let us now assume further that x^0 is an *interior* point, that is, $Ax^0 = b, x^0 > 0$.

One approach to defining a suitable local (reduced) model, when dropping inactive bounds, is to augment the objective function by a quadratic regularizing term, leading to the following local (reduced) model:

$$\text{minimize } c^T(x - x^0) + \frac{1}{2}(x - x^0)^T(x - x^0) + z^0$$
$$\text{s.t. } A(x - x^0) = 0. \tag{7.18b}$$

Let us now make the transformation of variables given by (7.3) and (7.5). Then from (7.18b), we derive the equivalent local (reduced) model,

$$\underset{\Delta x_N \in R^{n-m}}{\text{minimize}} (Z^T c)^T \Delta x_N + \frac{1}{2}\Delta x_N^T (Z^T Z)\Delta x_N + z^0. \tag{7.19}$$

The optimal point of this model is

$$\Delta x_N = -(Z^T Z)^{-1} Z^T c, \tag{7.20}$$

and the corresponding search direction in the original space, clearly a direction of descent, is

$$\Delta x = Z\Delta x_N = -Z(Z^T Z)^{-1} Z^T c \tag{7.21a}$$

$$= -[I - A^T(AA^T)^{-1}A]c. \tag{7.21b}$$

(The latter equality is a standard result concerning projectors. For an explicit proof, see Gill and Murray [18, p. 48].) Expression (7.21b) defines the direction used in a *range space* version of Rosen's method [49]. (Stating this in an alternative way, if we wished to use the projected-gradient direction at $x^0 > 0$, the local (reduced) model underlying this choice would be (7.18b).) Corresponding, (7.21a) is the direction that would be used in a *null space* version of Rosen's method.

A suitable step from x^0 along the direction Δx defined by (7.21a, b) can then be taken subject to satisfying the nonnegativity bounds, yielding an improving (interior) point. Its choice need not concern us for the moment as we are primarily concerned, in this section, with definition of search directions.

Since x^0 is an interior point, different basis matrices can be used to define (7.21a). Equality with (7.21b) shows that the search direction is *independent* of the choice of basis (up to permutation of variables).

To conclude this subsection we should mention, for completeness, that an alternative approach is to define the local (reduced) model by quadratic regularization applied to (7.4) instead of (7.1), that is, to make the quadratic regularization in the reduced space. With x^0 again an interior point, the corresponding local reduced model is

$$\underset{\Delta x_N \in R^{n-m}}{\text{minimize}} (Z^T c)^T \Delta x_N + \frac{1}{2}\Delta x_N^T \Delta x_N + z^0, \tag{7.22}$$

and from it we obtain

$$\Delta x_N = -Z^T c \quad \text{and} \quad \Delta x = -ZZ^T c. \tag{7.23}$$

This defines *Wolfe's reduced-gradient direction* (see Wolfe [60], where the reduced-gradient approach was first proposed; also see Murtagh and Saunders [37]) and is utilized by Murtagh and Saunders [38] when solving a linear program in reduced-gradient mode.

2.4. Projected and Reduced-Gradient Extensions with Transformation of Variables

In making the transition from (7.18a) to the local reduced model (7.18b), the latter remains unaware of how close the variables come to violating their bounds, and the direction deduced from it may lead to points that do so almost immediately. Since the components of x^0 are themselves a measure of how close bounds are to violation, let us instead employ in the quadratic regularization a metric that defines a measure of distance relative to x^0, to obtain the following model.

$$\begin{aligned} &\text{minimize } c^T(x - x^0) + \frac{1}{2}(x - x^0)^T D_0^{-2}(x - x^0) + z^0 \\ &\qquad \text{s.t. } A(x - x^0) = 0, \end{aligned} \tag{7.24}$$

where $D_0 \equiv \text{diag}[x_1^0, \ldots, x_n^0] > 0$. Since the quadratic term in (7.24) can be written as $\sum_{j=1}^n [(x_j - x_j^0)/x_j^0]^2$, we see that the metric employed is a measure of distance between x_j and x_j^0 taken *relative* to x_j^0, $1 \le j \le n$. It is also easily verified that (7.24) can be obtained, equivalently, by making the transformation of variables motivated by Karmarkar [28], namely $x = D_0\tilde{x}$, then employing a (Euclidean) quadratic regularization as in (7.18b), and finally reverting to the original variables.

The local reduced model deduced from (7.24) is thus

$$\underset{\Delta x_N \in R^{n-m}}{\text{minimize}} \ (Z^T c)^T \Delta x_N + \frac{1}{2}\Delta x_N^T Z^T D_0^{-2} Z \Delta x_N + z^0. \tag{7.25}$$

The minimizing point is

$$\Delta x_N = -(Z^T D_0^{-2} Z)^{-1} Z^T c \tag{7.26}$$

and the associated descent direction is

$$\Delta x = Z\Delta x_N = -Z(Z^T D_0^{-2} Z)^{-1} Z^T c \tag{7.27a}$$

$$= -D_0[I - D_0 A^T (A D_0^2 A^T)^{-1} A D_0] D_0 c. \tag{7.27b}$$

The last equality is also a standard result concerning projectors. It follows, for example, from a simple generalization of the proof, cited immediately after (7.21b), that replaces the Euclidean metric by the metric defined by an arbitrary positive definite symmetric matrix. There is clearly an analogy between (7.21a, b) and (7.27a, b). Expression (7.27b) defines the direction used in the *affine* (*scaling*) version of Karmarkar's method, and (7.27a) defines the direction that we shall use in a *null space affine* (*scaling*) version of Karmarkar's method. Again, as in Section 2.3, we have a direction that is independent of the choice of basis and we can take a step along it to a suitably improving (interior) point. Convergence of the procedure based on (7.27b) (and hence (7.27a)) under nondegeneracy assumptions and suitable choice of step has been established by Vanderbei et al. [59].

An important advantage of the form (7.27a) over (7.27b) is that the quantity Δx_N in (7.26) *can be approximated*, for example, by

$$v = -\Omega Z^T c, \tag{7.28}$$

where Ω is a positive definite symmetric matrix ($\Omega > 0$) that need *not* necessarily be known explicitly. The corresponding descent search direction $\Delta x = Zv$ continues to satisfy the equality constraints of (7.1) because $A\Delta x = AZv = 0$, using (7.2). In consequence, feasibility of a subsequent iterate can be preserved, *without incurring the expense of computing* Δx_N *exactly*. (Note also that analogous observations to the foregoing apply to (7.21a) vis-à-vis (7.21b).) However, we cannot directly deduce convergence of the resulting algorithm from the proof of Vanderbei et al. [59] and convergence must be established by a separate argument. We shall have more to say on approximation of the search direction later, in particular in Section 3.3.

§3. Algorithmic Considerations

Formulating (7.27a) into a mathematical algorithm requires further attention to detail. Specifically, we consider the following three points.

3.1 Choice of the Basis Matrix B Defining Z

Many linear programs have an embedded identity matrix (or some other matrix close to the identity that can be factorized and used to solve associated systems of linear equations with great efficiency). Because it is more convenient for present purposes, let us consider the symmetric primal-dual form of the linear programming problem in this and the next section, namely

$$\begin{aligned} \text{(P):} \quad & \text{minimize } c^T x \\ & \text{s.t. } Nx \geq b,\ x \geq 0 \end{aligned} \tag{7.29a}$$

and

$$\begin{aligned} \text{(D):} \quad & \text{maximize } b^T \pi \\ & \text{s.t. } N^T \pi \leq c,\ \pi \geq 0, \end{aligned} \tag{7.29b}$$

where N is an $m \times \bar{n}$ matrix of full rank. Introduce slack variables in the usual way into (P) to give the constraints $-v + Nx = b$, $v \geq 0$, $x \geq 0$, and define

$A = [-I \mid N]$ where A is an $m \times n$ matrix and $n \equiv m + \bar{n}$. The cost (row) vector associated with A is $[0 \mid c]^T$. When we choose the basis matrix in (7.5) to be the matrix $-I$ corresponding to the slack variables, we then have

$$Z = \begin{bmatrix} N \\ I_{\bar{n} \times \bar{n}} \end{bmatrix} \tag{7.30}$$

and

$$Z^T \begin{bmatrix} 0 \\ c \end{bmatrix} = c.$$

From (7.26) and (7.27a), the corresponding reduced gradient or null space affine (scaling) direction is given in the notation of (7.29) as follows.

$$\Delta x = -[N^T D_{v^0}^{-2} N + D_{x^0}^{-2}]^{-1} c, \quad \begin{bmatrix} \Delta v \\ \Delta x \end{bmatrix} = -\begin{bmatrix} N \\ I_{\bar{n} \times \bar{n}} \end{bmatrix} [N^T D_{v^0}^{-2} N + D_{x^0}^{-2}]^{-1} c, \tag{7.31}$$

where v^0, x^0 give an interior point of (P), $D_{v^0} \equiv \text{diag}[v_1^0, \ldots, v_m^0]$, $D_{x^0} \equiv \text{diag}[x_1^0, \ldots, x_{\bar{n}}^0]$, and D_0 is now the diagonal matrix with entries defined by D_{v^0} and D_{x^0}. Within the context of (7.29a), the direction used in Algorithm *I* of Adler et al. [1] obtained through a rather different derivation, is equivalent to (7.31). This is discussed in more detail in Section 6.3.

In an analogous manner, an identity matrix corresponding to slack variables, say w, can be appended to the constraints of (D), leading to a null space affine (scaling) direction for the dual as follows:

$$\begin{bmatrix} \Delta w \\ \Delta \pi \end{bmatrix} = -\begin{bmatrix} -N^T \\ I_{m \times m} \end{bmatrix} [N D_{w^0}^{-2} N^T + D_{\pi^0}^{-2}]^{-1} b. \tag{7.32}$$

3.2. Dimensionality of $[Z^T D_0^{-2} Z]$

For the primal (P) the matrix in (7.31) that must be factorized (inverted) is an $\bar{n} \times \bar{n}$ matrix, and for the dual (D) the corresponding matrix in (7.32) is an $m \times m$ matrix. A solution strategy that computes and factorizes these matrices (which we expressed more generally in Section 2 as $[Z^T D_0^{-2} Z]$), in order to solve the associated linear systems, would most likely opt to solve the primal when $\bar{n} \leq m$ and the dual when $\bar{n} > m$. However, when an iterative technique is used to solve the associated linear systems, the dimensionality of $[Z^T D_0^{-2} Z]$ can be misleading, as we now see in the next subsection.

3.3. Approximations to Δx_N

As we have noted, the null space affine (scaling) form has the significant advantage that it permits Δx_N to be approximated in (7.26) and (7.27a). This makes it especially advantageous to consider the use of iterative methods, in particular the method of conjugate gradients (or limited memory quasi-

Newton versions of it, see Nazareth [42]) to quickly obtain a good approximate solution. In contrast, (7.27b) requires accurate solution if feasibility is to be retained, and this makes direct methods more competitive for solving the associated linear systems.

Let us write the system defining Δx_N in (7.31), obtained from (7.26) for the particular choice of Z given by (7.30), as follows.

$$[N^T D_{v^0}^{-2} N + D_{x^0}^{-2}]\Delta x_N = -c. \tag{7.33}$$

Let us then reformulate it as

$$[D_{x_0} N^T D_{v^0}^{-2} N D_{x^0} + I]\Delta \hat{x}_N = -\hat{c}, \tag{7.34a}$$

where $\Delta \hat{x}_N \equiv D_{x^0}^{-1}\Delta x_N$, $\hat{c} \equiv D_{x^0} c$, and

$$\Delta x = \begin{bmatrix} N \\ I_{\bar{n}\times\bar{n}} \end{bmatrix} D_{x^0}\Delta\hat{x}_N. \tag{7.34b}$$

From (7.34a) and the definition $M \equiv D_{x^0} N^T D_{v^0}^{-2} N D_{x^0}$ we see that a *key* system of linear equations, which must be efficiently solved, is of the form

$$[I + M]\langle\text{variable}\rangle = \langle\text{right-hand-side}\rangle, \tag{7.35}$$

where M is a symmetric positive semidefinite matrix. The rank of M is obviously $l \equiv \min(m, \bar{n})$, and thus M has at most $l + 1$ distinct eigenvalues. Let λ_j denote the eigenvalues of M. The eigenvalues of $I + M$ are $1 + \lambda_j$ and the eigenvalues of M and $I + M$ obviously have the same multiplicities. Now it is well known that, in exact arithmetic, the CG method converges in as many steps as there are distinct eigenvalues of the system to which it is applied. By making identical arguments for the system involved in (7.32) (indeed by symmetry), we see that regardless of the relative sizes of m and $\bar{n}$, the *bound* on the number of CG iterations for the primal form (P) is the same as for the dual form (D), in exact arithmetic. Furthermore, M is positive semidefinite, so all eigenvalues of $I + M$ exceed unity. When the eigenvalues of M are not excessively large (large eigenvalues could occur in certain circumstances and necessitate refinements of the type discussed in Sections 5 and 6), we can expect the system (7.35) to be fairly well conditioned. Then the CG method will generally produce a reasonable approximation to the solution of (7.35) in a surprisingly low number of iterations, for example, the square root of the dimension, see Golub and Van Loan [23]. We can therefore expect to find that the problem (7.35) is amenable to solution by the conjugate gradient algorithm.

We note the following points concerning use of the conjugate gradient algorithm.

(i) The CG algorithm may be initiated at the origin and the matrix M is *not* formed explicitly.

(ii) The CG algorithm can be terminated before the exact solution to (7.35) is obtained. It is easy to verify by an inductive argument that such an inexact (or truncated) CG algorithm always produces a *descent* direction Δx. This may be deduced, for example, from results given in Hestenes [27]. Associated with the approximate solution of (7.35), say v, is a positive definite symmetric matrix Ω, which is not formed explicitly, with $v = -\Omega\hat{c}$, $\Delta x = Zv$. [See also (7.28).]

(iii) The transformation of variables used to define (7.34a) from (7.33) represents a particular *preconditioning* of the CG algorithm applied to (7.33).

The choice $B = I$ or $B = -I$ made possible by (7.29a, b) is exceedingly convenient. However, we *stress* the fact that the observations of Sections 3.2 and 3.3 hold for *any* basis matrix B in the more general setting (7.1) and (7.27a). We shall explore this further in Section 5.2. When B is carefully chosen, so that systems of equations of the form $B\langle\text{var}\rangle = \langle\text{rhs}\rangle$ or $B^T\langle\text{var}\rangle = \langle\text{rhs}\rangle$ can be solved with ease, this can lead to particularly effective computation of Δx in (7.27a) by means of expressions analogous to (7.31) through (7.35).

§4. A Numerical Experiment

In this section we study the *viability* of algorithms derived from (7.34a, b) within the context of a numerical experiment. Our overall approach to implementation is discussed in detail in Nazareth [39], where we describe the need to develop a hierarchy of implementations as an optimization method evolves from a mathematical algorithm (level 1) to reformulation as a numerical algorithm (level 2) to final implementation as good mathematical software (level 3). Here we are concerned with level-1 experiments on test problems of the symmetric primal-dual form, which happens to be more convenient than the standard form and is quite adequate for present needs.

We compare two algorithms derived from (7.34a, b) with the primal simplex algorithm and with one another. The algorithms derived from (7.34a, b) are as follows:

(1) RGK/LL^T

At any iteration, say k, this algorithm explicitly forms the matrix in (7.34a) and solves the associated system of linear equations by Cholesky factorization, forward elimination, and back substitution. The linear algebra algorithms used are given in Golub and Van Loan [23, pp. 89 and 53]. The search direction is then defined by (7.34b). The maximum step, say α, that can be taken along this direction without violating a nonnegativity bound is computed in the usual way and the step from the current to the next iterate is taken to be $0.9\alpha\Delta x$.

(2) RGK/CG

At any iteration, say k, this solves the system (7.34a) or (7.35) by carrying out up to a maximum of ncg_k steps of the conjugate gradient algorithm given by

Golub and Van Loan [23, p. 374], with no (further) preconditioning. (We refer the reader back to items (i) through (iii) of the discussion at the end of Section 3.3.) The maximum number of CG steps permitted at iteration k is controlled very simply as follows: the initial setting ncg_1 is taken to be the largest integer that does not exceed $m^{1/2}$ and thereafter $ncg_{k+1} \leftarrow ncg_k + k$. In addition, the conjugate gradient algorithm is terminated before ncg_k iterations are performed, if the residual for the linear system (7.35), say r, falls below some small quantity, as frequently happens. (Rather arbitrarily, we chose $r^T r \leq 10^{-3}$.) The computation of the step length is exactly as in the previous algorithm.

(3) $SMPLX/[I \mid B^{-1}N]$

This is the standard tableau form of the primal simplex algorithm [11], with one difference, namely that it incorporates the extension of the algorithm described in Section 2.1 for starting at an interior point. (Note that for dense problems with dimensions in (7.29a) satisfying $\bar{n} \leq 2m$, an iteration of the tableau simplex method is cheaper than an iteration of the revised simplex algorithm with dense basis matrix and complete pricing, see Chvatal [10, p. 114].) The initial basis is taken to be the columns corresponding to the slack variables, which is the same as the one chosen for the above RGK algorithms. *The difference is that thereafter, the basis is held fixed in* RGK/LL^T *and* RGK/CG, *but it is changed at each iteration of the simplex algorithm.*

We emphasize that the implementation of the above algorithms is almost a direct encoding of the underlying formulae, with the linear algebra being carried out without embellishment of the Golub-Van Loan procedures quoted above.

The test problems we use are given by Avis and Chvatal [3] (these being, in turn, patterned after Kuhn and Quandt [31]) and are of the form (7.29a), namely

$$\begin{aligned} &\text{minimize } -1^T x \\ &\text{s.t. } Nx \leq 10^4 1 \qquad (\text{equivalently } -Nx \geq -10^4 1), \qquad (7.36) \\ &x \geq 0. \end{aligned}$$

1 is a vector of all unit elements and N is a dense $m \times \bar{n}$ matrix with integer elements chosen at random in the range $1, \ldots, 1000$. (To completely specify the five test problems that we used requires that we identify the random number generator. We used the one supplied with Microsoft Basic Version 3.0. Random numbers are generated in the range $(0, 1]$ and are multiplied by 10^3 and rounded down to the nearest integer. Elements are generated by rows and the random number generator is, of course, reseeded before each test problem is generated.)

We avoid the problem of deciding whether or not (7.3b) should be dualized before applying algorithm RGK/LL^T (cf. Section 3.2) by taking $m = \bar{n}$. As in Avis and Chvatal [3], we choose $m = 10, 20, 30, 40, 50$. We also finesse the problem of finding a starting point, because one is readily available for the above family of test problems. We choose each component of the starting point x^0 to be $10/m$ and it can then be seen that the corresponding values of the slack variables are positive and the starting objective value is -10.0 independent of m. (For convenience, let us call this starting objective value z_{init}.) *We emphasize that for a given value of* m, *each algorithm is initiated with precisely the same starting point and run on precisely the same test problem of the form* (7.36).

The algorithms RGK/LL^T and RGK/CG generate an infinite sequence of iterates and are sensitive to stopping criteria. Because the primary purpose of our experimentation is to investigate *viability* of the RGK algorithms, we use the simplex algorithm as a benchmark and organize the experiment so as to circumvent the difficulty of making an appropriate choice of stopping criterion.

A choice is made for m $(= \bar{n})$. The simplex algorithm is run to completion with a stringent criterion based on (7.8) to ensure that optimality is attained. The objective values of the final iterates are examined and at the iterate, say K, where all figures up to the second (seems reasonable!) figure after the decimal point have settled down to their final optimal value, the run time, say T, is recorded. (Subsequent iterates are ignored.) In this way we avoid occasions, if any, where the simplex algorithm dawdles near its optimum, yet we also avoid premature termination. Let us denote the objective value, for iterate K, by z_{opt}. We then run algorithm RGK/LL^T until the completion of the *first* iteration, say J, for which the total run time exceeds T. The objective value reached is recorded and, if the run time is substantially in excess of T, the objective value obtained is averaged with the objective value at the end of the previous iteration (for which the run time is obviously less than T). Let us denote this objective value by z_f. In Table 7.1 we then report the relative error expressed as a percentage, namely

$$R = \frac{z_f - z_{\text{opt}}}{z_{\text{init}} - z_{\text{opt}}} 100\%. \tag{7.37}$$

Obviously this quantity is zero for the simplex algorithm since this algorithm is the benchmark. Table 7.1 also reports the number of iterations taken (upper left-hand corner of each entry), and if z_f is obtained by averaging as described above, the iteration count recorded is the fraction $(J + (J - 1))/2$.

The entire procedure is repeated for algorithm RGK/CG. In addition to the above information, Table 7.1 reports the average number of CG steps taken (total number of CG steps divided by number of iterations taken by RGK/CG) and the maximum and minimum number of CG steps at any iteration. In Table 7.1 they are given in the last column, in the order maximum, average, minimum. (Also, for $m = 10$ we set $ncg_1 \leftarrow 1$.)

The algorithms were implemented in Microsoft Basic (the iteractive features of the interpreter make up, to some extent, for the shortcomings of the language) and run on an IBM PC/AT-compatible, in double-precision arithmetic.

Table 7.1

Dimension	$SMPLX[I:B^{-1}N]$ iteration count	$SMPLX[I:B^{-1}N]$ R	RGK/LL^T iteration count	RGK/LL^T R	RGK/CG iteration count	RGK/CG R	RGK/CG max. CG steps	RGK/CG average CG steps	RGK/CG min. CG steps
10	12	0.	2.5	17.8	3.5	11.52	6	3.3	1
20	30	0.	3	7.4	5	1.44	13	7.6	3
30	37	0.	2	28.0	5	2.6	14	8	3
40	85	0.	3.5	6.46	7.5	0.67	21	12.6	3
50	128	0.	4	3.2	10	0.48	27	14.6	3

Each entry in the foregoing table is as follows:

iteration count		max. CG steps
	R see (7.37)	average CG steps
		min. CG steps

When evaluating a mathematical algorithm by relating the performance of an experimental implementation on a relatively small number of sample problems (level 1), the fact that test results *have* been obtained and that they are reasonable is often of more significance than the particular numbers that are tabulated. In other words, the main value of such results lies in the fact that they help one to discern certain broad trends in performance and to confirm the theoretical soundness of the underlying method. Our testing has been very much in this vein. In particular, in this section we sought to design *an unbiased numerical experiment* in order to evaluate and compare different algorithmic possibilities arising from the discussion of Sections 2.4 and 3 and to guide our further investigation in subsequent sections of this chapter. We should also note that although we use a random number generator to generate our five test problems (each of different dimension), ours is *not* a Monte Carlo simulation study. Reservations have been expressed about the Kuhn-Quandt problems when they are used in a statistical experiment. These do not apply to our simple but, we believe, instructive numerical experiment summarized in Table 7.1.

Our limited test results reported in Table 7.1 and the "feel" that develops from fairly extensive interactive experimentation during the process of implementing the algorithms and generating the results lead us to the conclusion that algorithm RGK/LL^T performs surprisingly well and that an advanced implementation of algorithm RGK/CG is a promising route to pursue. There is much scope for fine tuning of the CG procedure and for the use of CG variants to utilize information from prior iterations, in particular to enhance efficiency during the final stages of algorithm *RGK/CG*. We must be careful about making extrapolations to the large, sparse case, but the CG method is certainly most at home in this setting. On the other hand, there are devices for enhancing efficiency of algorithms based on (7.27b) that take advantage of the fixed sparsity structure (cf. Karmarkar [29]), and in this context some interesting ideas for enhancing efficiency by dropping elements of A in (7.1) that are below a certain threshold, coupled with iterative refinement, have been investigated by Parlett [45]. Such ideas would carry across to RGK/LL^T algorithms.

Finally, we have observed that there is also scope for "hybridizing" *RGK*-type algorithms and that the primal simplex algorithm, as formulated in Section 2.1, can pick up *at any interior point*, say x_f, where an *RGK* algorithm leaves off. If the latter is used to identify a good basis, the primal simplex algorithm should very quickly find the optimal solution when initiated from x_f. In such circumstances our experimentation has indicated that most iterations involve moving nonbasic variables to a bound without the need for any basis change. At the same time the apparatus of the simplex method is available for identifying columns that must be included in the optimal basis.

§5. Combining Models

We have just noted that it is desirable to have a simplex-like algorithm as a core procedure to drive an interior point to a vertex, for example, a point where a null space affine (scaling) algorithm terminates. There are several additional reasons for augmenting an algorithm based on an affine (scaling) model of the form (7.25) by also using a simplex-like model of the form (7.7) in an appropriate way.

(a) In a column generation procedure, for example, a procedure based on the Dantzig-Wolfe decomposition principle [12], all the columns of the LP matrix are obviously not available simultaneously for use in the model (7.25).
(b) A starting point may be available that is *on* a facet of the feasible polytope.
(c) An affine (scaling) algorithm may generate a point near (and thereafter hug) the boundary of the feasible polytope, see Megiddo and Shub [34]. (Clearly such observations also apply to the null space affine (scaling) algorithm.) This has a detrimental effect both on the total iteration count and on the conditioning of the linear systems that define search directions.

The symmetric primal-dual form (7.29) was convenient for the purpose of level-1 experiments in Sections 3 and 4. However, let us now revert to the standard form (7.1) of Section 2, because it is more commonly used in practice. In order to achieve the augmentation just discussed, we follow Murtagh and

Saunders [37, 38] and further partition the nonbasic columns as

$$N = [S \mid T]. \tag{7.38}$$

The matrix S denotes the columns corresponding to the s *superbasic* variables x_S, that is, nonbasic variables whose corresponding bounds are slack. The matrix T denotes the columns corresponding to the remaining t nonbasic variables x_T that are fixed (or tight), usually (though not necessarily) at their bounds. Obviously $s + t = n - m$. (Since variables can be reordered, there is no loss of generality in assuming that superbasics precede other nonbasics.)

We partition the matrix Z, the vector x, and the step $\Delta x \equiv x - x^0$ as follows.

$$Z \equiv [Z_S \mid Z_T] \equiv \begin{bmatrix} -B^{-1}S & -B^{-1}T \\ I_{s\times s} & 0 \\ 0 & I_{t\times t} \end{bmatrix}, \quad x = \begin{bmatrix} x_B \\ x_S \\ x_T \end{bmatrix}, \quad \Delta x \equiv \begin{bmatrix} \Delta x_B \\ \Delta x_S \\ \Delta x_T \end{bmatrix}. \tag{7.39}$$

5.1. The Relaxing Step

The step Δx_T in the space of the corresponding variables x_T is obtained from a model derived from (7.7), namely

$$\begin{aligned} &\text{minimize } (Z_T^T c)^T \Delta x_T + z^0 \\ &\text{s.t. } x_T^0 + \Delta x_T \geq 0. \end{aligned} \tag{7.40}$$

The simplex algorithm [11] chooses a single coordinate, for example, the one with the most negative reduced cost, and the corresponding column of Z_T defines the direction of search. If we now make it a policy to change as *many* coordinates simultaneously as possible, a natural choice is to select the subset, say $\tilde{T}$, corresponding to all coordinates in the reduced space with negative reduced costs, that is, $\tilde{T} \equiv \{ j \mid (Z_T^T c)_j < 0\}$, and to define the search direction $\Delta x = \sum_{j \in \tilde{T}} |\sigma_j| z_j$; in this expression z_j is the jth column of Z_T, and it is weighted by the absolute value of $\sigma_j \equiv (Z_T^T c)_j = z_j^T c$. This is equivalent to defining Δx_T as the *negative projected reduced gradient*, namely

$$(\Delta x_T)_j \equiv \max[0, -(Z_T^T c)_j], \qquad j = 1, \ldots, t, \tag{7.41}$$

and choosing the direction of search that corresponds to it, namely

$$\Delta x = Z_T \Delta x_T. \tag{7.42}$$

5.2. The Restricted Step

The step Δx_S in the space of the superbasic variables x_S is obtained from a model of the form (7.25), namely

$$\text{minimize } (Z_S^T c)^T \Delta x_S + \tfrac{1}{2} \Delta x_S^T W \Delta x_S + z^0 \tag{7.43}$$

where

$$W = [S^T B^{-T} D_B^{-2} B^{-1} S + D_S^{-2}], \tag{7.44}$$

and $D_B = \text{diag}[x_1^0, \ldots, x_m^0] > 0$, $D_S = \text{diag}[x_{m+1}^0, \ldots, x_{m+s}^0] > 0$. The search direction in the restricted space is obtained by solving the system

$$[S^T B^{-T} D_B^{-2} B^{-1} S + D_S^{-2}]\Delta x_S = -Z_S^T c.$$

For reasons discussed in Section 3.3, the conjugate gradient algorithm can be used to solve a reexpression of this system in the form

$$[I + M_S]\Delta \hat{x}_S = -\hat{c}, \tag{7.45}$$

where $\Delta \hat{x}_S = D_S^{-1} \Delta x_S$, $\hat{c} \equiv D_S Z_S^T c$, and $M_S \equiv D_S S^T B^{-T} D_B^{-2} B^{-1} S D_S$.

An algorithm that combines relaxing and restricted steps within the algorithmic framework of Dembo [13] and Dembo and Sahi [14] is given in Nazareth [43] and is currently under investigation.

§6. Practical Aspects

To conclude, let us now turn briefly to the more general linear program

$$\begin{aligned} &\text{minimize } c^T x \\ &\text{s.t. } Ax = b, \; l \leq x \leq u, \end{aligned} \tag{7.46a}$$

which is the form best suited to practical computation. Here l and u denote vectors of lower and upper bounds, respectively.

6.1. Bounds

Let us assume that all bounds are finite. It is simple matter to convert the linear program to the following form, where for convenience we use the same notation as (7.46a).

$$\begin{aligned} &\text{minimize } c^T x \\ &\text{s.t. } Ax = b, \; x + w = u, \; x \geq 0, \; w \geq 0. \end{aligned} \tag{7.46b}$$

Then given a feasible interior point $x^0 > 0$, $w^0 > 0$, the local reduced model analogous to (7.25) is

$$\text{minimize } (Z^T c)^T \Delta x_N + \tfrac{1}{2} \Delta x_N^T (Z^T (D_0^{-2} + W_0^{-2}) Z) \Delta x_N + z^0, \tag{7.47}$$

where $D_0 = \text{diag}[x_1^0, \ldots, x_n^0]$ and $W_0 = \text{diag}[w_1^0, \ldots, w_n^0]$. Analogous expressions to (7.27a, b) can be derived for the resulting search direction.

The corresponding generalization of (7.7), and also of (7.40), (7.43), (7.44), and (7.45), is straightforward.

6.2. Dual Variables

Under the assumption that x^0 is an interior point so that the bound constraints of (7.1) can be locally dropped, recall the local (reduced) model (7.24), namely

$$\begin{aligned} &\text{minimize } c^T(x - x^0) + \tfrac{1}{2}(x - x^0)^T D_0^{-2}(x - x^0) + z^0 \\ &\text{s.t. } A(x - x^0) = 0. \end{aligned} \tag{7.48}$$

Second-order estimates of Lagrange multipliers π of (7.48) can immediately be obtained from its Lagrangian conditions as follows:

$$A^T\pi = D_0^{-2}(x - x^0) + c.$$

Using $A(x - x^0) = 0$ implies that

$$(AD_0^2A^T)\pi = AD_0^2c.$$

Hence

$$\pi = (AD_0^2A^T)^{-1}AD_0^2c. \tag{7.49}$$

Similarly, for problem (7.46) we could use

$$\pi = (A(D_0^2 + W_0^2)A^T)^{-1}A(D_0^2 + W_0^2)c. \tag{7.50}$$

6.3. Correspondences and Their Practical Implications

Next, we discuss correspondences with the affine (scaling) algorithm described in Adler et al. [1] and related algorithms described by Gay [17] and Monma and Morton [36].

Consider first the linear program (7.29a) and reexpress it in the inequality form used in Adler et al. [1], namely

$$\text{maximize } -c^Tx \quad \text{s.t.} \begin{bmatrix} -N \\ -I_{\bar{n}\times\bar{n}} \end{bmatrix} x \leq \begin{bmatrix} -b \\ 0 \end{bmatrix}. \tag{7.51}$$

Introduce slack variables v_1 and v_2 into (7.51), that is,

$$\begin{gathered} \text{maximize } (-c)^Tx \\ \text{s.t.} \begin{bmatrix} -N \\ -I_{\bar{n}\times\bar{n}} \end{bmatrix} x + \begin{bmatrix} I_{m\times m} & \cdot \\ \cdot & I_{\bar{n}\times\bar{n}} \end{bmatrix} \begin{bmatrix} v_1 \\ v_2 \end{bmatrix} = \begin{bmatrix} -b \\ 0 \end{bmatrix}, \\ v_1 \geq 0,\ v_2 \geq 0. \end{gathered} \tag{7.52}$$

At a given feasible interior point x^0, with associated values $v_1^0 > 0$ and $v_2^0 > 0$ for the slack variables, the affine (scaling) direction of Adler et al. [1] can immediately be stated as follows. See, in particular, their expression (2.17).

$$\begin{bmatrix} d_{v_1} \\ d_{v_2} \end{bmatrix} = -\begin{bmatrix} -N \\ -I_{\bar{n}\times\bar{n}} \end{bmatrix} \left[[-N^T \mid -I_{\bar{n}\times\bar{n}}] \begin{bmatrix} D_{v_1^0}^{-2} & \cdot \\ \cdot & D_{v_2^0}^{-2} \end{bmatrix} \begin{bmatrix} -N \\ -I_{\bar{n}\times\bar{n}} \end{bmatrix} \right]^{-1} (-c).$$

Use $x = v_2$, $x^0 = v_2^0$ in (7.52), define $v \equiv v_1$, $v^0 \equiv v_1^0$, and simplify the foregoing expression. Then this direction is as follows.

$$\begin{bmatrix} d_v \\ d_x \end{bmatrix} = -\begin{bmatrix} N \\ I_{\bar{n}\times\bar{n}} \end{bmatrix} [N^T D_{v^0}^{-2} N + D_{x^0}^{-2}]^{-1} c. \tag{7.53}$$

We see that (7.53) and the second expression of (7.31) are identical.

Conversely, consider a linear program in the canonical form employed by Adler et al. [1].

$$\text{maximize } c^T x \quad \text{s.t. } Ax \leq b. \tag{7.54}$$

The foregoing program is the dual of a linear program stated in standard form. Therefore affine (scaling) algorithms based on it are often termed *dual* affine algorithms. Introduce slack variables v, namely

$$\text{maximize } c^T x \quad \text{s.t. } v + Ax = b,\ v \geq 0. \tag{7.55}$$

At a given feasible interior point x^0 with associated values $v^0 > 0$ for the slack variables, the affine (scaling) direction of Adler et al. [1] is as follows. See expressions (2.16) and (2.17) of their paper.

$$\begin{bmatrix} d_v \\ d_x \end{bmatrix} = \begin{bmatrix} -A \\ I \end{bmatrix} [A^T D_{v^0}^{-2} A]^{-1} c. \tag{7.56}$$

A null space affine (scaling) direction for the linear program (7.55) at v^0, x^0 would utilize the following local reduced model function.

$$\begin{aligned} &\text{minimize } [0 \mid -c]^T \begin{bmatrix} v - v^0 \\ x - x^0 \end{bmatrix} + \frac{1}{2}(v - v^0)^T D_{v^0}^{-2} (v - v^0) \\ &\text{s.t. } [I \mid A] \begin{bmatrix} v - v^0 \\ x - x^0 \end{bmatrix} = 0. \end{aligned} \tag{7.57}$$

Here the x variables are omitted from the quadratic term in the objective function because these variables have no associated bounds. The null space of $[I \mid A]$ is spanned by $\begin{bmatrix} -A \\ I \end{bmatrix}$, and the null space affine (scaling) direction is given by a derivation completely analogous to (7.31), that is,

$$\begin{bmatrix} d_v \\ d_x \end{bmatrix} = -\begin{bmatrix} -A \\ I \end{bmatrix} [A^T D_{v^0}^{-2} A]^{-1}(-c) = \begin{bmatrix} -A \\ I \end{bmatrix} [A^T D_{v^0}^{-2} A]^{-1} c. \tag{7.58}$$

Directions (7.56) and (7.58) are identical.

Adler et al. [1] and Monma and Morton [36] point out that the inequality form (7.54) allows inexact projections without loss of *feasibility*. However, this is not sufficient. If the resulting search direction is to be a direction of *descent*, it is equally important to emphasize that the approximation should be explicitly (or *implicitly*) of the form $-Hc$, where H is a positive Semidefinite symmetric matrix that is related to $\Omega > 0$ in expression (7.28), by $H = Z\Omega Z^T$.

This is true, for example, when an iterative method like the conjugate gradient algorithm is used to solve the system of linear equations associated with (7.56) or (7.58), even when the algorithm is terminated after just a few steps. *We believe that the implementations given in Adler et al.* [1] *and Momma and Morton* [36] *can be considerably accelerated by inexact computation of search directions along lines discussed in this chapter.*

When a linear program is in standard form, an identity basis matrix may not be available. In the null space affine approach, a basis matrix other than the identity matrix can be utilized to define the search direction. Recall, in particular, expression (7.45). However, the matrix that defines the search direction may then be a completely dense matrix. To maintain it as the product of sparse matrices, it would be imperative to use an *iterative* method to solve the associated system of linear equations.

Finally, note that the null space approach is well established in the field of nonlinear programming. Many of the implementation techniques developed in this more general setting can be suitably adapted to linear programming. See Nazareth [43] for a further discussion.

§7. Concluding Remarks

In their pure forms, the primal simplex algorithm and Karmarkar's projective (scaling) algorithm differ substantially in both their overall philosophy (active set strategy as compared to treating the entire problem simultaneously) and their properties (relatively weak invariance properties as compared to fairly strong invariance properties, worst-case exponential under various pricing strategies as compared to worst-case polynomial (bit) complexity). Experimental evidence, however, reveals one important point in common, namely that *on average* both the simplex algorithm and Karmarkar's projective (scaling) algorithm take a remarkably low number of *iterations* to reach a close neighborhood of the solution of a linear program. This observation seemed to suggest that there may be underlying connections and motivated our earlier work on an affine (scaling) version derived via a homotopy approach, Nazareth [40], which in turn led us to study the null space affine (scaling) version discussed here.

It has been the purpose of this paper to argue that *from the standpoint of implementation* the simplex method and Karmarkar's method can draw on one another in a substantial manner, leading to algorithmic variants that complement one another nicely. The discussion of Sections 2 and 3, the numerical experimentation of Section 4, and the discussion and algorithm of Sections 5 and 6 are intended to support this point of view; namely, in a practical implementation, the simplex method can provide the core algorithm (because it deals so effectively with "warm" starts and because it almost literally embodies duality theory and the LP optimality criterion), yet there is considerable room for maneuver in augmenting the core procedure in a rather natural manner, using an algorithm based on the null space affine (scaling) approach. In this way one can deal effectively with situations where less information is available about a starting solution ("cold" starts). A harmonious blending of the two approaches seems desirable through the use of suitable adaptive strategies that are governed by the particular characteristics of the problem being solved. Indeed, most large-scale mathematical programming systems utilize such adaptive strategies; see, for example, Benichou et al. [7].

Acknowledgments

My thanks to Professor Beresford Parlett for helpful comments on this chapter. Some useful suggestions of the referee are also gratefully acknowledged.

References

[1] I. Adler, N. Karmarkar, M. G. C. Resende, and G. Veiga, An implementation of Karmarkar's algorithm for linear programming, Report No. ORC 86-8, Operations Research Center, University of California, Berkeley (1986, revised 1987).

[2] K. Anstreicher, A monotonic projective algorithm for fractional linear programming, Preprint, Yale School of Organization and Management, Yale University, New Haven, Conn. (1985).

[3] D. Avis and V. Chvatal, Notes on Bland's pivoting rule, *Math. Programming Study* **8** (1978), 24–34.

[4] E. R. Barnes, A variation of Karmarkar's algorithm for solving linear programming problems, Research Report No. 11136, IBM T. J. Watson Research Center, Yorktown Heights, N.Y. (1985).

[5] E. R. Barnes, S. Chopra, and D. Jensen, Polynomial-time convergence of the affine scaling algorithm with centering, Presented at the Conference on Progress in Mathematical Programming, Asilomar Conference Center, Pacific Grove, Calif. (March 1–4, 1987).

[6] D. A. Bayer and J. C. Lagarias, The nonlinear geometry of linear programming. I: Affine and projective scaling trajectories, Preprint, AT&T Bell Laboratories, Murray Hill, N.J. (1986).

[7] M. Benichou, J. M. Gauthier, G. Hentges, and G. Ribiere, The efficient solution of large-scale linear programming problems—some algorithmic techniques and computational results, *Math. Programming* **13** (1977), 280–232.

[8] L. Blum, A new simple homotopy algorithm for linear programming, Preprint, Department of Mathematics, University of California, Berkeley (1987).

[9] V. Chandru and B. Kochar, A class of algorithms for linear programming, Preprint, Department of Industrial Engineering, Purdue University, Lafayette, Ind. (1985).

[10] V. Chvatal, *Linear Programming*, Freeman, San Francisco (1983).

[11] G. B. Dantzig, *Linear Programming and Extensions*, Princeton University Press, Princeton, N.J. (1963).

[12] G. B. Dantzig, and P. Wolfe, Decomposition principle for linear programs, *Operations Res.* **8** (1960), 101–111.

[13] R. Dembo, A primal truncated Newton algorithm with application to large-

scale nonlinear network optimization, Technical Report No. YALEU/DCS/TR304, Department of Computer Science, Yale University, New Haven, Conn. (also Yale School of Organization and Management, Working Paper Series B#72).

[14] R. Dembo and S. Sahi, A convergent framework for constrained nonlinear optimization, Technical Report No. YALEU/DCS/TR303, Department of Computer Science, Yale University, New Haven, Conn. (also Yale School of Organization and Management, Working Paper Series B#69).

[15] R. Dembo and T. Steihaug, Truncated-Newton algorithms for large-scale unconstrained optimization, *Math. Programming* **26** (1983), 190–212.

[16] D. Gay, A variant of Karmarkar's linear programming algorithm for problems in standard form, Numerical Analysis Manuscript 85–10, AT&T Bell Laboratories, Murray Hill, N.J. (1985).

[17] D. Gay, Pictures of Karmarkar's linear programming algorithm, CS Technical Report No. 136, AT&T Bell Laboratories, Murray Hill, N.J. (1987).

[18] P. E. Gill and W. Murray, *Numerical Methods for Constrained Optimization*, Academic Press, London, 1974.

[19] P. E. Gill, W. Murray, M. A. Saunders, J. A. Tomlin, and M. H. Wright, On projected Newton barrier methods for linear programming and an equivalence to Karmarkar's projective method, Technical Report SOL 85–11, Systems Optimization Laboratory, Department of Operations Research, Stanford University, Stanford, Calif. (1985).

[20] D. Goldfarb and S. Mehrotra, A relaxed version of Karmarkar's method, Technical Report, Department of Industrial Engineering and Operations Research, Columbia University, New York (1985).

[21] D. Goldfarb and J. K. Reid, A practicable steepest-edge simplex algorithm, *Math. Programming* **12** (1977), 361–371.

[22] E. G. Golshtein, An iterative linear programming algorithm based on an augmented Lagrangian, in *Nonlinear Programming*, vol. 4, O. L. Mangasarian, R. R. Meyer and S. M. Robinson (eds.), Academic Press, New York, 1981, pp. 131–146.

[23] G. H. Golub and C. F. Van Loan, *Matrix Computations*, John Hopkins University Press, Baltimore, 1983.

[24] H. Greenberg, Pivot selection tactics, in *Design and Implementation of Optimization Software*, H. Greenberg (ed.), NATO Advanced Studies Institute Series E, Applied Science, No. 28, Sijthoff and Noordhoff, 1978, pp. 143–174.

[25] H. Greenberg and J. Kalan, An exact update for Harris' TREAD, *Math. Programming Study* **4** (1975), 26–29.

[26] P. M. J. Harris, Pivot selection methods in the Devex LP code, *Math. Programming Study* **4** (1975), 30–57.

[27] M. R. Hestenes, *Conjugate Direction Methods in Optimization*, Springer-Verlag, New York, 1980.

[28] N. Karmarkar, A new polynomial-time algorithm for linear programming, *Combinatorica* **4** (1984), 373–395.

[29] N. Karmarkar, Lecture at the International Congress of Mathematicians, University of California, Berkeley (August 1986).

[30] L. G. Khachiyan, A polynomial algorithm for linear programming, *Dokl. Akad. Nauk SSSR* **244** (1979), S, 1093–1096; translated in *Soviet Math. Doklady* **20** (1979), 191–194.

[31] H. Kuhn and R. E. Quandt, An experimental study of the simplex algorithm, in N. C. Metropolis et al. (eds.), *Experimental Arithmetic, High-speed Computing and Mathematics*, Proceedings of Symposium on Applied Mathematics XV, American Mathematical Society, Providence, R.I., 1963, pp. 107–124.

[32] N. Megiddo, On the complexity of linear programming, in *Advances in Economic Theory the Fifth World Congress*, T. Bewlen (ed.), Cambridge University Press, 1987, pp. 225–226.

[33] N. Megiddo, Pathways to the optimal set in linear programming. In: *Progress in Mathematical Programming: Interior-Point and Related Methods*. Springer-Verlag, New York (1989).

[34] N. Megiddo and M. Shub, Boundary behaviour of interior point algorithms in linear programming, *Mathematics of Operations Research* (to appear).

[35] O. L. Mangasarian, Iterative solution of linear programs, *SIAM J. Numer. Anal.* **18** (1981), 606–614.

[36] C. L. Monma and A. J. Morton, Computational experience with a dual affine variant of Karmarkar's method for linear programming, Preprint, Bell Communications Research, Morristown, N.J. (1987).

[37] B. A. Murtagh and M. A. Saunders, Large-scale linearly constrained optimization, *Math. Programming* **14** (1978), 41–72.

[38] B. A. Murtagh and M. A. Saunders, MINOS 5.0 User's Guide, Technical Report SOL 83-20, Systems Optimization Laboratory, Department of Operations Research, Stanford University, Stanford, Calif. (1983).

[39] J. L. Nazareth, Hierarchical implementation of optimization methods, in P. Boggs, R. Byrd, and B. Schnabel (eds.), *Numerical Optimization, 1984*, SIAM, Philadelphia, 1985, pp. 199–210.

[40] J. L. Nazareth, Homotopy techniques in linear programming *Algorithmica* **1** (1986), 529–535.

[41] J. L. Nazareth, Implementation aids for optimization algorithms that solve sequences of linear programs, *ACM Trans. Math. Software* **12** (1986), 307–323.

[42] J. L. Nazareth, Conjugate gradient methods less dependent on conjugacy, *SIAM Rev.* **28** (1986), 501–511.

[43] J. L. Nazareth, Pricing criteria in linear programming, Report PAM-382, Center for Pure and Applied Mathematics, University of California, Berkeley (1987).

[44] J. L. Nazareth, *Computer Solution of Linear Programs*, Oxford University Press, Oxford and New York, 1987.

[45] B. N. Parlett, private communication (1986).

[46] B. N. Pschenichny and Y. M. Danilin, *Numerical Methods in Extremum Problems*, M.I.R. Publishers, Moscow, 1975; English translation, 1978.

[47] J. Renegar, A polynomial-time algorithm, based on Newton's method, for linear programming, Report 07118-86, Mathematical Sciences Research Institute, Berkeley, Calif. (1986).

[48] R. T. Rockafellar, Augmented Lagrangians and applications of the proximal point algorithm in convex programming, *Math. Operations Res.* **1** (1976), 97–116.

[49] J. B. Rosen, The gradient projection method for non-linear programming. Part I: Linear constraints", *SIAM J. Appl. Math.* **8** (1960), 181–217.

[50] D. F. Shanno, A reduced gradient variant of Karmarkar's algorithm, Working Paper 85-01, Graduate School of Administration, University of California, Davis (1985) (revised version coauthored with R. Marsten).

[51] N. Z. Shor, Utilization of space dilation operation in minimization of convex functions, *Kibernetika* **1** (1970), 6–12.

[52] N. Z. Shor, Generalized gradient methods of non-differentiable optimization employing space dilation operations, in A. Bachem, M. Grotschel, and B. Korte (eds.), *Mathematical Programming: The State of the Art*, Springer-Verlag, Berlin, 1983, pp. 530–539.

[53] S. Smale, Algorithms for solving equations, Preprint of lecture at the International Congress of Mathematicians, University of California, Berkeley (August 1986).

[54] D. Solow, *Linear Programming: An Introduction to Finite Improvement Algorithms*, North-Holland, New York and Amsterdam, 1984.

[55] G. Strang, Karmarkar's algorithm in a nutshell, *SIAM News* **18** (1985), 13.

[56] M. J. Todd and B. P. Burrell, An extension of Karmarkar's algorithm for linear programming using dual variables, Technical Report 648, School of Operations Research and Industrial Engineering, Cornell University, Ithaca, N.Y. (1985).

[57] K. Tone, A hybrid method for linear programming, Report 85-B-1, Graduate School of Policy Science, Saitama University, Urawa, Saitama, Japan (1985).

[58] P. M. Vaidya, An algorithm for linear programming which requires $O(m + n)n^2 + (m + n)^{1.5}n)L)$ arithmetic operations, Preprint, AT&T Bell Laboratories, Murray Hill, N.J. (1987).

[59] R. J. Vanderbei, M. J. Meketon, and B. A. Freedman, A modification of Karmarkar's linear programming algorithm, Preprint, AT&T Bell Laboratories, Holmdel, N.J. (1985).

[60] P. Wolfe, The reduced-gradient method, unpublished manuscript, Rand Corporation, Santa Monica, Calif. (1962).

[61] Y. Ye, Karmarkar-type algorithm for nonlinear programming, Preprint, Engineering-Economics Systems Department, Stanford University, Stanford, Calif. (1986).

[62] W. I. Zangwill, *Nonlinear Programming*, Prentice-Hall, Englewood Cliffs, N.J., 1969.

CHAPTER 8

Pathways to the Optimal Set in Linear Programming

Nimrod Megiddo

Abstract. This chapter presents continuous paths leading to the set of optimal solutions of a linear programming problem. These paths are derived from the weighted logarithmic barrier function. The defining equations are bilinear and have some nice primal-dual symmetry properties. Extensions to the general linear complementarity problem are indicated.

§1. Introduction

Algorithms for mathematical programming can often be interpreted as path-following procedures. This interpretation applies to the simplex method [4], Scarf's fixed-point algorithm [19], Lemke's algorithm [13] for the linear complementarity problem, homotopy methods for piecewise linear equations [5], and most of the methods for nonlinear optimization. This is the theme of the book by Garcia and Zangwill [8]. More recent algorithms for linear programming by Murty [17] and Mangasarian [15] are also based on natural paths that lead to optimal solutions. Iterative algorithms for nonlinear optimization usually assign to any point x in a certain set $S \subset R^n$ (usually convex) a "next point" $x' = f(x) \in S$. Given a starting point x^0, the iterative scheme generates a sequence of points $\{x^k\}$, where $x^{k+1} = f(x^k)$, that converges to a solution.

It is often instructive to consider "infinitesimal" versions of iterative algorithms in the following sense. Given the iterative scheme $x' = f(x)$, consider the differential equation

$$\dot{x} = f(x) - x.$$

This chapter was published in the Proceedings of the 7th Mathematical Programming Symposium of Japan, Nagoya, Japan, pp. 1–35, and is referenced by several other papers in this volume. Since those Proceedings are not highly available, it is reproduced here in its original form.

When this equation has a unique solution through x^0 then it determines a path $x = x(t)$ such that the tangent to the path at any x is equal to the straight line determined by x and x'. If the algorithm generates the point x' close to x then the path may be a good approximation to the sequence generated by the algorithm. This is true at least during later stages of the execution if the sequence converges to a solution point. If the algorithm makes large steps during the early stages then the path may be a bad approximation. Trajectories corresponding to discrete algorithms for nonlinear optimization were analyzed in [6, 8, 10]. The analogy to differential equations is well known.

Several people have recently worked on solution paths in linear programming. Nazareth [18] interprets Karmarkar's algorithm [12] as a homotopy method with restarts. Results about the infinitesimal version of Karmarkar's algorithm and related algorithms were recently obtained in [2] and [16]. Smale [20] showed that the path generated by the self-dual simplex algorithm [4] can be approximated by the Newton's method path for solving a certain system of nonlinear equations.

In this chapter we study solution paths related to barrier functions for linear programming. We believe the study of paths is essential for the design and analysis of algorithms for optimization. In Section 2 we describe the paths for linear programming problems in standard form. In Section 3 we develop essentially the same theory within a more symmetric framework. In Section 4 we analyze some properties of tangents to the trajectories, whereas Section 5 brings some observations on higher-order derivatives. In Section 6 we consider the behavior of trajectories near corners. In Section 7 we discuss generalizations to the linear complementarity problem.

§2. On the Logarithmic Barrier Function

In this section we consider the linear programming problem in the standard form

$$\text{(P)} \qquad \begin{aligned} &\text{Maximize } c^T x \\ &\text{subject to } Ax = b, \\ &\qquad\qquad\quad x \geq 0 \end{aligned}$$

where $A \in R^{m \times n}$, $b \in R^m$, and $c, x \in R^n$. We believe most of the readers are used to considering the linear programming problem in this form. However, an analogous analysis can be carried out with respect to other forms of the problem. We shall later discuss the problem in more detail, using another more symmetric variant. The presentation will therefore entail a fair amount of redundancy, which, we hope, will be of help to the reader.

The logarithmic barrier function technique, usually used in nonlinear constrained optimization, can of course be applied to the linear programming problem. This method recently came up in [9], where Karmarkar's algorithm

[12] was analyzed from the barrier function viewpoint, but the idea of using this function in the context of linear programming is usually attributed to Frisch [7]. The technique gives rise to the following problem:

$$(\mathrm{P}_\mu) \qquad \begin{aligned} &\text{Maximize } c^T x + \mu \sum_j \ln x_j \\ &\text{subject to } Ax = b, \\ &\qquad\qquad x > 0 \end{aligned}$$

where $\mu > 0$ is typically small. The barrier function approach is valid only if there exists an $x > 0$ such that $Ax = b$. However, it is easy to reformulate the problem, using one artificial variable, so that the feasible domain is of full dimension. We use e to denote a vector of 1's of any dimension as required by the context. Also, M denotes a real number always chosen to be sufficiently large or an "infinite" element adjoined to the ordered field of the reals. The following construction is well known. Given a problem in the form (P), consider the following problem:

$$(\mathrm{P}^*) \qquad \begin{aligned} &\text{Maximize } c^T x - M\xi \\ &\text{subject to } Ax + (b - Ae)\xi = b, \\ &\qquad\qquad x, \xi \geq 0. \end{aligned}$$

Obviously, x is an optimal solution for (P) if and only if $(x, 0)$ is an optimal solution for (P*). It follows that the vector $(x, \xi) = e$ satisfies the set of equations. Thus, without loss of generality, we may assume the problem is given in the form (P) and also $Ae = b$.

For any d-vector x, let D_x denote the diagonal matrix of order $d \times d$ whose diagonal entries are the components of x. A vector $x > 0$ is an optimal solution for (P_μ) if and only if there exists a vector $y \in R^m$ such that

$$(\mathrm{O}) \qquad \begin{aligned} \mu D_x^{-1} e \quad - A^T y &= -c, \\ Ax \qquad\qquad &= b. \end{aligned}$$

Obviously, the problem (P_μ) may be unbounded. Let us assume, for a moment, that the feasible domain $\{x\colon Ax = b, x \geq 0\}$ is bounded. At least in this case both (P) and (P_μ) have optimal solutions (for every μ). Under the boundedness assumption, (P_μ) has a *unique* optimal solution for every $\mu > 0$ since its objective function is strictly concave. Thus, under the boundedness assumption, the system (O) has a unique solution for x for every $\mu > 0$.

The left-hand side of the system (O) represents a nonlinear mapping $F_\mu(x, y)$ of R^{n+m} into itself. The Jacobian matrix of this mapping at (x, y) is obviously the following:

$$J = J_\mu(x, y) = \begin{pmatrix} -\mu D_x^{-2} & -A^T \\ A & 0 \end{pmatrix}.$$

Suppose A is of full rank m $(m \leq n)$. In this case, the value of y is uniquely determined by the value of x. Also, it is well known that in this case the matrix AA^T is positive definite and hence nonsingular. The linear system of equations

$$J_\mu(x, y)\begin{bmatrix} \xi \\ \eta \end{bmatrix} = \begin{bmatrix} \mu \\ 0 \end{bmatrix}$$

can be interpreted as a "least squares" problem or a projection problem. A solution can be expressed in terms of the matrix $(AD_x^2A^T)^{-1}$, which is well defined since A is of full rank. It is interesting to observe the following:

Proposition 8.1. *The problem* (P_μ) *is either unbounded for every* $\mu > 0$ *or has a unique optimal solution for every* $\mu > 0$.

PROOF: Consider the interval I of values t for which the set

$$L(t) = \{x \geq 0 : c^T x = t, Ax = b\}$$

has a nonempty interior. Obviously, I is an open interval. If for any $t \in I$ the function $\phi(x) = \sum_j \ln x_j$ is unbounded on $L(t)$ then, of course, (P_μ) is unbounded for all positive values of μ. Without loss of generality, assume $\phi(x)$ is bounded over every $L(t)$ $(t \in I)$. Strict concavity of $\phi(x)$ implies that for each $t \in I$ there is a unique maximizer $x = x(t)$ of ϕ over $L(t)$. Let $g(t)$ denote the maximum value of $\phi(x)$ over $L(t)$. Consider first the case where $\infty \in I$; that is, the function $c^T x$ is unbounded. Here there is a *ray*, contained in the interior of the feasible region, along which $c^T x$ tends to infinity. Since the domain is polyhedral, the ray is bounded away from the boundary. Thus, on the ray the function $\phi(x)$ is bounded from below, and hence (P_μ) is unbounded for every $\mu > 0$. In the remaining case, notice that strict concavity of $g(t)$ implies that $t + \mu g(t)$ is bounded for every $\mu > 0$ if t is bounded. Thus, in the latter case (P_μ) has a unique optimal solution for every $\mu > 0$.

It follows from Proposition 8.1 that if the system (O) has a solution for any positive value of μ then it determines a unique and continuous path $x = x(\mu)$, where μ varies over the positive reals. When A is of full rank also a continuous path $y = y(\mu)$ is determined. We are interested in the limits of $x(\mu)$ and $y(\mu)$ as μ tends to zero. Suppose (for a moment) that the limits of $x(\mu)$ and $y(\mu)$ (as μ tends to 0) exist, and denote them by $\bar{x}$ and $\bar{y}$, respectively. It follows that $A\bar{x} = b, \bar{x} \geq 0$, and $A^T\bar{y} \geq c$. Moreover, for each j such that $\bar{x}_j > 0$, $A_j^T\bar{y} = c_j$. It follows that $\bar{x}$ and $\bar{y}$ are optimal solutions for (P) and its dual, respectively. To relate these paths to an algorithm for the linear programming problem, we have to address at least two issues. First, we have to know a solution for, say, $\mu = 1$. Second, the limit of $x(\mu)$ (as μ tends to zero) should exist.

It is easy to modify the objective function so that an initial solution becomes available. Note that instead of (P_μ) we can work with a problem of the form

$$(P_\mu(w)) \qquad \begin{aligned} &\text{Maximize } c^T x + \mu \sum_j w_j \ln x_j \\ &\text{subject to } Ax = b, \\ &\qquad\qquad x > 0, \end{aligned}$$

where $w \in R_+^n$ is any vector with positive components. Proposition 8.1 extends to this case. Suppose x^0 and y^0 are interior feasible solutions for the primal and the dual problems, respectively. We will show later that any problem can be reformulated so that such solutions are readily available. We can choose w so that the vectors x^0 and y^0 satisfy the optimality conditions with respect to $(\mathrm{P}_\mu(w))$ at $\mu = 1$:

$$(\mathrm{P}(w)) \qquad \begin{aligned} \mu D_x^{-1} w \quad - A^T y &= -c \\ Ax \qquad\qquad &= b. \end{aligned}$$

Specifically, $w = D_{x^0}(A^T y^0 - c)$. Thus, given any pair of interior feasible solutions for the primal and the dual problems, we can easily calculate a suitable weight vector w, which in turn determines paths $x = x(\mu)$ and $y = y(\mu)$ as explained above. We discuss the role of the weights in more detail in Section 3.

In view of the preceding discussion, let us assume that for every $\mu > 0$ the system (O) has a unique solution $(x(\mu), y(\mu))$. It is easy to show that $c^T x(\mu)$ tends to the optimal value of (P). This follows if we multiply the first row of (O) by $x(\mu)$, the second by $y(\mu)$, and then add them up. We get $b^T y(\mu) - c^T x(\mu) = n\mu$. The optimal value lies between $b^T y(\mu)$ and $c^T x(\mu)$ and this implies our claim that $c^T x(\mu)$ tends to the optimal value as μ tends to 0. We are interested in conditions under which the *point* $x(\mu)$ tends to an optimal solution of (P).

Let $V(\mu) = c^T x(\mu)$ (where $x(\mu)$ is the optimal solution of (P_μ)), and let $V(0)$ denote the optimal value of (P). We have just argued that $V(\mu)$ tends to $V(0)$ as μ tends to 0. Obviously, $x(\mu)$ is also the optimal solution of the following problem:

$$(\bar{\mathrm{P}}_\mu) \qquad \begin{aligned} &\text{Maximize } c^T x + \mu \sum_j \ln x_j \\ &\text{subject to } Ax = b, \\ &\qquad\qquad c^T x = V(\mu), \\ &\qquad\qquad x > 0. \end{aligned}$$

The latter is of course equivalent to

$$\begin{aligned} &\text{Maximize } \sum_j \ln x_j \\ &\text{subject to } Ax = b, \\ &\qquad\qquad c^T x = V(\mu), \\ &\qquad\qquad x > 0. \end{aligned}$$

Our assumption of existence of the path $x(\mu)$ is equivalent to existence of an optimal solution for the problem $(\bar{\mathrm{P}}_\mu)$ for any $\mu > 0$. Using the notation of Proposition 8.1, the function $\phi(x)$ is bounded on every $L(t)$ where $t = V(\mu)$ for some $\mu > 0$. We assert that this implies that the set $L(t)$ itself is bounded. The proof is as follows. If $L(t)$ is unbounded then there is a ray, bounded away from the boundary, along which at least one of the variables tends to infinity while the others are bounded away from zero. Along such a ray the function $\phi(x)$ tends to infinity. It follows that the set $L(V(0))$ is bounded. The maximum value $g(t)$ is a concave function of t. This concavity implies that $g(t)$ is bounded from above as t tends to $V(0)$. Let N denote the set of all indices j such that $x_j = 0$ in *every* optimal solution. Thus, the optimal face is the intersection of the feasible domain with subspace $\{x: x_j = 0, j \in N\}$. Let

$$\phi_N(x) = \sum_{j \in N} \ln x_j$$

and

$$\phi_B(x) = \phi(x) - \phi_N(x).$$

Let $\xi_j(\mu)$ denote the jth component of the vector $x(\mu)$, that is, the optimal solution at μ. Since $\phi_N(x)$ is constant on the set $\{x: x_j = \xi_j(\mu), j \in N\}$, it follows that the point $x(\mu)$ is actually the optimal solution of the problem

$$\begin{aligned} &\text{Maximize } \phi_B(x) \\ &\text{subject to } Ax = b, \\ &\qquad c^T x = V(\mu), \\ &\qquad x_j = \xi_j(\mu) \qquad (j \in N), \\ &\qquad x > 0. \end{aligned}$$

Since the optimal set is bounded, it follows that the problem corresponding to $\mu = 0$, that is,

$$\begin{aligned} &\text{Maximize } \phi_B(x) \\ &\text{subject to } Ax = b, \\ &\qquad c^T x = V(0), \\ &\qquad x_j = 0 \qquad (j \in N), \\ &\qquad x_j > 0 \qquad (j \notin N), \end{aligned}$$

has a unique optimal solution which we denote by $x(0)$. We claim that $x(0)$ is equal to the limit of $x(\mu)$ as μ tends to zero. This solution is also characterized by the following system:

$$\begin{aligned} &\frac{1}{x_j} - A_j^T y = -\lambda c_j \qquad (j \notin N), \\ &Ax = b, \\ &x_j = 0 \qquad (j \in N), \\ &c^T x = V(0), \end{aligned}$$

where λ is a multiplier corresponding to the equation $c^T x = V(0)$ and A_j^T is the jth row of A^T. Any limit of a convergent sequence of points $x(\mu_k)$ (where μ_k tends to 0 as k tends to infinity) satisfies the latter system of equations and hence equals $x(0)$. Thus, $x(\mu)$ tends to $x(0)$ as μ tends to zero. We can thus state the following proposition:

Proposition 8.2. *If for some $\mu > 0$ the system* (O) *has a solution $x > 0$ then for every $\mu > 0$ there is a solution $x(\mu)$ so that the path $x(\mu)$ is continuous and the limit of $x(\mu)$ as μ tends to zero exists and constitutes an optimal solution to the linear programming problem* (P).

The implication of Proposition 8.2 is that we can solve the linear programming problem by a "homotopy" approach. Starting from $\mu = 1$, where we readily have an optimal solution to problems of the form $(P_\mu(w))$, we follow the path of optimal solutions for such problems while μ varies from 1 to 0. The limit as μ tends to zero is an optimal solution to the linear programming problem, namely the point $x(0)$. In the next section we will continue the study of the paths introduced above. However, henceforth we will consider a more symmetric form of the problem.

§3. Duality

We find it more instructive to consider the linear programming problem in the *symmetric* form (in the sense of the duality transformation):

$$\text{(P)}\qquad \begin{aligned} &\text{Maximize } c^T x \\ &\text{subject to } Ax \le b, \\ &\qquad\qquad x \ge 0 \end{aligned}$$

where $A \in R^{m \times n}$, $b \in R^m$, and $c, x \in R^n$. The system $\{Ax \le b\}$ can obviously be replaced by $\{Ax + u = b, u \ge 0\}$ where $u \in R^m$. In this section we complement the results of Section 2 and provide additional insights.

The following nonlinear concave optimization problem (where μ is a fixed positive number) can be considered an approximation to (P):

$$(\text{P}_\mu)\qquad \begin{aligned} &\text{Maximize } c^T x + \mu\left(\sum_j \ln x_j + \sum_i \ln u_i\right) \\ &\text{subject to } Ax + u = b, \\ &\qquad\qquad x, u > 0. \end{aligned}$$

Notice that the gradient of the function $\phi(x) = \sum_j \ln x_j$ is equal to $D_x^{-1}e$ and also to $D_x^{-2}x$. A pair of vectors $x \in R^n_+$ and $u \in R^m_+$, such that $Ax + u = b$, constitutes an optimal solution for (P_μ) if and only if there exists a vector $y \in R^m$ such that

$$\begin{aligned} \mu D_x^{-2} x \quad & - A^T y = -c, \\ \mu D_u^{-1} e \quad & -y \;\;= \;\; 0. \end{aligned}$$

It follows that such a vector y must satisfy $u = \mu D_y^{-1} e$ and hence x is optimal in (P_μ) if and only if there is $y \in R_+^m$ such that

$$\text{(O)} \qquad \begin{aligned} \mu D_x^{-2} x \quad & - A^T y \;\;= -c, \\ Ax \quad & + \mu D_y^{-2} y = \;\; b. \end{aligned}$$

The system (O) has some nice symmetry properties. Consider the dual of (P), namely

$$\text{(D)} \qquad \begin{aligned} &\text{Minimize } b^T y \\ &\text{subject to } A^T y \geq c, \\ &\qquad\qquad y \geq 0. \end{aligned}$$

An approximate nonlinear convex optimization problem is as follows.

$$(\mathrm{D}_\mu) \qquad \begin{aligned} &\text{Minimize } b^T y - \mu\left(\sum_i \ln y_i + \sum_j \ln v_j\right) \\ &\text{subject to } A^T y - v = c, \\ &\qquad\qquad y, v > 0. \end{aligned}$$

It is easy to check that y is optimal in (D_μ) if and only if there exists an $x \in R_+^n$ such that (O) holds. Note that the nonlinear objective functions of (P_μ) and (D_μ) (for $\mu > 0$) are strictly concave and strictly convex, respectively. Thus, each of the problems (P_μ) and (D_μ) has at most one optimal solution. The relationship between the problems (P_μ) and (D_μ) is summarized in the following duality theorem:

Theorem 8.1.

(i) *If the problem* (P_μ) *is unbounded then the problem* (D_μ) *is infeasible, and if the problem* (D_μ) *is unbounded then the problem* (P_μ) *is infeasible.*
(ii) *The problem* (P_μ) *has a optimal solution if and only if the problem* (D_μ) *has an optimal solution.*
(iii) *If x and y are optimal solutions for* (P_μ) *and* (D_μ)*, respectively, then the gap between the values of the objective functions of* (P_μ) *and* (D_μ) *is equal to* $(m + n)\mu(1 + \ln \mu)$*, whereas the gap between* $c^T x$ *and* $b^T y$ *equals* $(m + n)\mu$*.*

PROOF. Suppose x and y are feasible solutions to the problems (P_μ) and (D_μ), respectively. Let u and v be as above. It follows that

$$\begin{aligned} y^T A x + u^T y &= b^T y, \\ y^T A x - v^T x &= c^T x. \end{aligned}$$

Thus, for $\mu \le 1$,

$$b^T y - c^T x = u^T y + v^T x \ge \mu(u^T y + v^T x)$$

$$> \mu\left(\sum_i \ln u_i + \sum_i \ln y_i + \sum_j \ln v_j + \sum_j \ln x_j\right).$$

Thus,

$$b^T y - \mu\left(\sum_i \ln y_i + \sum_j \ln v_j\right) > c^T x + \mu\left(\sum_j \ln x_j + \sum_i \ln u_i\right).$$

The latter is analogous to the weak duality in linear programming. It implies that if (P_μ) is unbounded then (D_μ) is infeasible and if (D_μ) is unbounded then (P_μ) is infeasible.

We know from the preceding discussion of the system (O) that (P_μ) has an optimal solution if and only if (D_μ) has one. The optimal solutions are unique. If x and y are the optimal solutions for (P_μ) and (D_μ), respectively, then the system (O) implies

$$v_j x_j = u_i y_i = \mu.$$

Thus,

$$b^T y - c^T x = (m + n)\mu$$

and

$$\left[b^T y - \mu\left(\sum_i \ln y_i + \sum_j \ln v_j\right)\right] - \left[c^T x + \mu\left(\sum_j \ln x_j + \sum_i \ln u_i\right)\right]$$

$$= (m + n)\mu(1 - \ln \mu).$$

Interestingly, the gap between the optimal values depends only on μ and the dimensions m and n and not on the data. It follows from Theorem 8.1 that the optimal solutions $x = x(\mu)$ and $y = y(\mu)$ are such that $c^T x(\mu)$ and $b^T y(\mu)$ tend to the optimal value of (P) (which of course equals the optimal value of (D)). Moreover, the "duality gap" tends to zero *linearly* with the parameter μ. It can then be shown, as in the preceding section, that the points themselves tend to optimal solutions of (P) and (D), respectively.

For the symmetric primal-dual barrier approach to work, we need both (P) and (D) to have full-dimensional feasible domains. We note that every linear programming problem can be reformulated so that both the primal and the dual have full-dimensional feasible domains. Given a problem in the form (P), consider the following problem, where M is sufficiently large:

$$(\mathrm{P}^*) \qquad \begin{aligned} &\text{Maximize } c^T x - M\xi \\ &\text{subject to } Ax + (b - Ae - e)\xi \le b, \\ &\qquad (c - A^T e + e)^T x \le M, \\ &\qquad x, \xi \ge 0. \end{aligned}$$

It is easy to verify that if M is sufficiently large then x is an optimal solution for (P) if and only if $(x, 0)$ is an optimal solution for (P*). The point $e \in R^{n+1}$ lies in the interior of the feasible domain of (P*). Also, the point $e \in R^{m+1}$ lies in the interior of the feasible domain of the dual of (P*):

$$\text{(D*)} \qquad \begin{aligned} &\text{Minimize } b^T y + M\eta \\ &\text{subject to } A^T y + (c - A^T e + e)\eta \geq c, \\ &\qquad (b - Ae - e)^T y \geq -M, \\ &\qquad y, \eta \geq 0. \end{aligned}$$

Tricks of "Big M" are fairly standard in linear programming. Alternatively, to avoid numerical problems with large values of M, we can use here the equivalent of what is called "Phase I" in the linear programming literature.

For simplicity of notation, we write (x, y) for the column vector obtained by concatenating two column vectors x and y. We find it interesting to consider the mapping $\psi: R^{n+m} \to R^{n+m}$, defined by

$$\psi(x, y) = (\mu D_x^{-2} x - A^T y, Ax + \mu D_y^{-2} y).$$

This mapping underlies the system (O), which can be written as $\psi(x, y) = (-c, b)$. The Jacobian matrix of ψ at (x, y) is equal to

$$H = \begin{pmatrix} -\mu D_x^{-2} & -A^T \\ A & -\mu D_y^{-2} \end{pmatrix}.$$

Assuming x and y are positive, the matrix H is negative definite since for any $w \in R^n$ and $z \in R^m$,

$$(z, w)^T H (z, w) = -\mu(z^T D_x^{-2} z + w^T D_y^{-2} w).$$

In particular, H is nonsingular. It is also interesting to consider a related symmetric matrix

$$\tilde{H} = \begin{pmatrix} -\mu D_x^{-2} & -A^T \\ -A & +\mu D_y^{-2} \end{pmatrix}.$$

Obviously, $\tilde{H}$ is the Hessian matrix of the function

$$L_\mu(x, y) = c^T x + \mu \sum_j \ln x_j - y^T A x - \mu \sum_i \ln y_i + y^T b,$$

which is well defined for $x, y > 0$. Note that L is strictly concave in x for every y and strictly convex in y for every x. The pair $(x(\mu), y(\mu))$ (that is, the point where the gradient of $L(x, y)$ vanishes) constitutes the unique saddle point of $L(x, y)$, in the sense that x is a maximum and y is a minimum.

The sum of logarithms added to the linear objective function $c^T x$ plays the role of a "barrier" [6]. Suppose an algorithm for unconstrained optimization starts in the interior of the feasible domain and iterates by searching a line through the current point. The barrier "forces" the iterates to remain in the interior of the feasible domain. Another classical trick of nonlinear programming is to use a "penalty" function (where a penalty is incurred if a point outside the feasible domain is produced). Let us consider general algorithms that iterate on primal and dual interior points. Let μ denote a parameter that determines primal and dual interior feasible solutions $x(\mu) > 0$ and $y(\mu) > 0$, respectively. Let

$$u(\mu) = b - Ax(\mu) > 0$$

and

$$v(\mu) = A^T y(\mu) - c > 0.$$

If $x(\mu)$ and $y(\mu)$ tend (as μ tends to 0) to optimal solutions of the primal and dual problems, respectively, then necessarily the products $x_j(\mu)v_j(\mu)$ and $y_i(\mu)u_i(\mu)$ tend to 0 with μ. In other words, there exist functions $\mu_i(\mu)$ and $v_j(\mu)$ that tend to zero with μ so that

$$b_i - A_i x = \mu_i(\mu)\frac{1}{y_i}$$

and

$$A_j^T y - c_j = v_j(\mu)\frac{1}{x_j}.$$

The logarithmic barrier function method with uniform weights is *characterized* by the equations

$$\mu_i(\mu) = v_j(\mu) = \mu.$$

With general (not necessarily uniform) weights the functions μ_i and v_j remain linear in μ.

In pursuit of "natural" barrier or penalty functions, let us consider a problem in the following general form:

$$(\mathrm{P}_{f,\mu}) \qquad \begin{aligned} &\text{Maximize } c^T x + \mu \sum_j f(x_j) + \mu \sum_i f(u_i) \\ &\text{subject to } Ax + u = b \end{aligned}$$

where $f(\xi)$ is strictly concave. Let $g(\xi) = f'(\xi)$ and for any d-vector a let

$$G_a = \mathrm{Diag}(g(a_1), \ldots, g(a_d)).$$

A pair (x, u) is optimal for $(\mathrm{P}_{f,\mu})$ if and only if there exists a vector $y \in R^m$ such that

$$\begin{aligned} \mu G_x e \quad - A^T y &= -c, \\ \mu G_u e \quad\quad - y &= 0. \end{aligned}$$

We would like to have optimality conditions that are "primal-dual symmetric," that is, similar to the system (O) above. More precisely, we are interested in functions $f(\xi)$ where the optimal solution for the approximate dual problem provides Lagrange multipliers supporting the optimal solution of the approximate primal, and vice versa. Such functions would give rise to duality theorems similar to Theorem 8.1. When u is eliminated by the substitution $u_i = g^{-1}(y_i/\mu)$, we obtain a set of equations that we would like to have the same form as $\mu G_x e - A^T y = -c$. In other words, we need the function $g(\xi)$ to satisfy

$$\mu g(\xi) = g^{-1}\left(\frac{\xi}{\mu}\right)$$

for every ξ and $\mu > 0$. The latter requirement is very restrictive. It implies $g(\xi) = g^{-1}(\xi)$ so that $\mu g(\xi) = g(\xi/\mu)$. It follows that $g(\xi) = g(1)/\xi$. We reach the surprising conclusion that the only barrier or penalty functions that are primal-dual symmetric are of the form $f(\xi) = \kappa \ln(|\xi|)$, where κ is some constant. Such functions are appropriate only as barrier functions, that is, for interior point procedures, and not as penalty functions (for exterior point procedures).

We have already argued that for any pair (x^0, y^0) of interior feasible solutions (for (P) and (D), respectively), there exist weights that determine a pair of weighted barrier paths from x^0 and y^0 to the optimal sets. The characterization of these paths is simple. For simplicity of notation, let the indices of columns and rows vary over disjoint sets so that we can use w_i to denote a weight associated with a row and w_j to denote one associated with a column. Given the interior points x^0 and y^0, let

$$w_j = [A_j^T y^0 - c_j] x_j^0$$

and

$$w_i = [b_i - A_i x^0] y_i^0.$$

Then the function

$$c^T x + \mu\left(\sum_j w_j \ln x_j + \sum_i w_i \ln u_i\right)$$

has a maximum over the interior of the primal feasible region. Also, the function

$$b^T y - \mu\left(\sum_i w_i \ln y_i + \sum_j w_j \ln v_j\right)$$

has a minimum over the interior of the dual feasible region. If W is the total of the weights then the gap between the values of the linear functions is equal to $W\mu$. The gap between the values of the nonlinear functions is equal to $W\mu(1 - \ln \mu)$. The paths are characterized by the property that along each of them the products of complementary variables $x_j v_j$ and $y_i u_i$ are proportional to μ. In other words, the ratios across these products are kept constant. More explicitly,

$$x_j v_j = \mu w_j$$

and

$$y_i u_i = \mu w_i$$

along the paths. In the following section we will study the differential descriptions of these paths.

We have argued that a solution path determined by a pair of interior feasible solutions (for the primal and the dual problems) is the locus of interior feasible points with the same ratios across products of complementary variables. This interpretation suggests a natural generalization. Consider the following set of equations:

$$\text{(X)} \qquad \begin{aligned} x_j(A_j^T y - c_j) &= \mu w_j, \\ y_i(b_i - A_i x) &= \mu w_i. \end{aligned}$$

The original problem (P) requires that $Ax < b$ and $x \geq 0$. However, we can consider 2^{n+m} different problems, corresponding to the 2^{n+m} different ways of choosing the restrictions on the signs of the variables x_j and $u_i = b_i - A_i x$. The dual problem to each of these is obtained by suitable changes of sign of the complementary dual variables. For all such pairs of primal and dual problems, the products of complementary variables have to be nonnegative. In other words, if all the products $x_j v_j$ and $y_i u_i$ are nonnegative, then x and y are (respectively) primal and dual feasible solutions for at least one of these pairs of problems. In any case, x and y are feasible solutions of some pair of problems (not necessarily dual) that can be obtained from the original ones by changing the directions of some inequalities. The system (X) defines solution paths for all the feasible combinations of primal and dual problems.

A convenient description of the paths discussed above is obtained as follows. First, consider the problem

$$(\mathrm{P}_\infty) \qquad \begin{aligned} &\text{Maximize } \sum_j \ln x_j + \sum_i \ln u_i \\ &\text{subject to } Ax + u = b, \\ &\qquad\qquad x, u > 0. \end{aligned}$$

which is, in a sense, the limit of (P_μ) as μ tends to infinity. If (P_∞) has an optimal solution x^∞ then $x(\mu)$ tends to x^∞ as μ tends to infinity. Second, consider the problem of minimizing

$$c^T x - \mu\left(\sum_j \ln x_j + \sum_i \ln u_i\right)$$

subject to the same constraints. It is easy to see that as μ tends to infinity the path of the latter also approaches x^∞. It seems nice to apply at this point a change of parameter so that the paths of the two optimization problems can be described in a unified way. Consider the substitution $\mu = \tan\theta$. Equivalently, consider maximizing the following nonlinear objective function:

$$(\cos\theta)c^Tx - (\sin\theta)\left(\sum_j \ln x_j + \sum_i \ln u_i\right).$$

For $0 < \theta < \pi/2$ we get the part of the path corresponding to the minimization problem, whereas the interval $\pi/2 < \theta < \pi$ corresponds to the maximization problem. The value $\theta = \pi/2$ corresponds to maximization of the sum of logarithms. If the intersections of level sets of c^Tx with the feasible polyhedron are bounded and the linear problem has a minimum then the path is well defined for $0 \leq \theta < \pi/2$. If the feasible polyhedron is unbounded then the path is not defined at $\theta = \pi/2$. In fact, it diverges to infinity as θ tends to $\pi/2$. The defining equations of the path have the form

$$\begin{aligned} (\sin\theta)D_x^{-1} \qquad & -A^Ty &= -(\cos\theta)c, \\ Ax \qquad & +(\sin\theta)D_y^{-1} &= \ (\cos\theta)b. \end{aligned}$$

Again, if the domain is bounded, this system defines a continuous path that leads from a minimum of c^Tx to a maximum of c^Tx through the maximum of the sum of logarithms.

It is interesting to consider the system discussed above in the neighborhood of $\theta = 0$. We know that the limits $x(0)$ and $y(0)$ exist (if the paths exist). We first prove.

Proposition 8.3. *Let $\bar{x}$, $\bar{y}$, $\bar{u}$, and $\bar{v}$ denote the optimal values of variables in* (P_μ) *and* (D_μ) *at the end of the paths (that is, when μ tends to zero, assuming the problem has an optimal solution). Then, for each pair of complementary variables, $(\bar{x}_i, \bar{u}_i)$ and $(\bar{y}_j, \bar{v}_j)$, one member of the pair is positive while the other equals zero.*

PROOF. Obviously, at least one of the members in each pair equals zero. It is well known that at degenerate vertices some pairs may have both members equal zero. However, degeneracy means that either the primal or the dual problem has an optimal face of dimension greater than zero. We claim that the solution paths converge to points in the relative interior of the optimal faces of the primal and dual problems. Consider, for example, the primal problem. The limit point $\bar{x}$ is where the sum $\sum_j \ln x_j + \sum_i \ln u_i$ (taken over all the variables that are not identically zero on the optimal face) is maximized relative to the optimal face. Obviously, each variable that is not identical to zero on the optimal face does not vanish at $\bar{x}$. This implies our proposition.

Assuming the limits $x(0)$ and $y(0)$ exist, consider the variables that vanish at this point. They also vanish at every other point of the optimal set. Let I denote the set of indices i such that $A_i x = b_i$ at every primal optimal solution x. Also, let J denote the set of indices j such that $x_j = 0$ at every primal optimal solution x. It follows that for every dual optimal solution y, $y_i = 0$ for $i \notin I$ and $A_j^T y = c_j$ for every $j \notin J$. Consider the following problem:

$$
(\tilde{\mathrm{P}}) \quad
\begin{aligned}
&\text{Minimize } c^T x \\
&\text{subject to } A_i x \ge b_i \quad (i \in I), \\
&\qquad\qquad A_i x \le b_i \quad (i \notin I), \\
&\qquad\qquad x_j \le 0 \quad (j \in J), \\
&\qquad\qquad x_j \ge 0 \quad (j \notin J).
\end{aligned}
$$

This problem is approximated by

$$
(\tilde{\mathrm{P}}_\mu) \quad
\begin{aligned}
&\text{Minimize } c^T x - \mu\left(\sum_{j \in J} \ln(-x_j) + \sum_{j \notin J} \ln x_j + \sum_{i \in I} \ln(-u_i) + \sum_{i \notin I} \ln u_i\right) \\
&\text{subject to } Ax + u = b, \\
&\qquad x_j < 0\ (j \in J),\ x_j > 0\ (j \notin J),\ u_i < 0\ (i \in I),\ u_i > 0\ (i \notin I).
\end{aligned}
$$

It follows that the optimality conditions for $(\tilde{\mathrm{P}}_\mu)$ are the same as those for (P_μ) in the sense that the solution paths (assuming they exist on both sides) can be joined continuously at the optimal face common to problems (P) and $(\tilde{\mathrm{P}})$. Recall that the function $c^T x$ increases monotonically as μ tends to zero. It follows that, as long as the path can be continued, it can be extended through the hyperplane arrangement so that in every cell it travels (monotonically in terms of $c^T x$) from a minimum of the cell to a maximum of the cell, which is also a minimum of an adjacent cell, then to a maximum of this adjacent cell, and so on. The substitution $\mu = \tan\theta$ yields a continuous representation of a combined path that travels through a sequence of bounded cells. Each sequence of bounded cells can be extended on both sides with unbounded cells where the path tends to infinity. Except in pathological cases, the paths do not visit cells in which the function has neither a maximum nor a minimum. A pathological case is, for example, a polyhedral cylinder on which the linear function is unbounded (both from above and from below).

§4. On Tangents to the Paths

Let (x^0, y^0) be a pair of interior feasible solutions (for problems (P) and (D), respectively). Let u^0 and v^0 denote the corresponding slack vectors of the primal and dual problems, respectively. We use the products $w_j = x_j^0 v_j^0$ and $w_i = y_i^0 u_i^0$ to define a pair of paths as explained earlier. Let us now examine the *tangent* to this path at the starting point.

The path is determined by the following equations:

$$
\begin{aligned}
\mu w_j \frac{1}{x_j} \quad - A_j^T y &= -c_j, \\
A_i x \quad + \mu w_i \frac{1}{y_i} &= b_i.
\end{aligned}
$$

Differentiation with respect to μ yields the following equations:

$$-\mu w_j \frac{\dot{x}_j}{x_j^2} \quad - A_j^T \dot{y} = -w_j \frac{1}{x_j},$$

$$A_i \dot{x} \quad -\mu w_i \frac{\dot{y}_i}{y_i^2} = -w_i \frac{1}{y_i}.$$

Consider a point (x', y') with $x' = x^0 - \delta \dot{x}$ and $y' = y^0 - \delta \dot{y}$, where $\dot{x}$ and $\dot{y}$ constitute the solution of the latter system of equations at $x = x^0$, $y = y^0$, and $\mu = 1$ and δ is any positive number. Obviously, (x', y') lies on the tangent to the curve at x^0 and y^0. It is easy to verify that

$$b^T y' - c^T x' = (1 - \delta)(b^T y - c^T x).$$

Let us denote the slack vectors corresponding to the pair (x', y') by u' and v', and let w_i' and w_j' denote the corresponding products of complementary variables. It follows that

$$\begin{aligned} w_j' &= x_j' v_j' \\ &= (x_j^0 - \delta \dot{x}_j)(v_j^0 - \delta A_j^T \dot{y}) \\ &= w_j \left[1 - \delta + \delta^2 \frac{\dot{x}_j}{x_j^0} \left(1 - \frac{\dot{x}_j}{x_j^0} \right) \right]. \end{aligned}$$

Similarly,

$$w_i' = y_i' u_i' = w_i \left[1 - \delta + \delta^2 \frac{\dot{y}_i}{y_i^0} \left(1 - \frac{\dot{y}_i}{y_i^0} \right) \right].$$

For the points x' and y' to remain feasible in their respective problems, it is necessary and sufficient that the following quantities be less than or equal to 1:

$$\delta \frac{\dot{x}_j}{x_j^0}, \qquad \delta \left(1 - \frac{\dot{x}_j}{x_j^0} \right), \qquad \delta \frac{\dot{y}_i}{y_i^0}, \qquad \delta \left(1 - \frac{\dot{y}_i}{y_i^0} \right).$$

It is interesting to examine properties of the tangents. To establish some connections to other interior point methods, we return for a moment to the problem in standard form as in Section 2. Thus, we now work with the problem in the form

$$\begin{aligned} &\text{Maximize } c^T x \\ \text{(P)} \qquad &\text{subject to } Ax = b, \\ &\qquad x \geq 0. \end{aligned}$$

Given a pair (x^0, y^0) where

$$Ax^0 = b, \qquad x^0 > 0 \quad \text{and} \quad A^T y^0 > c,$$

the pair of paths $x = x(\mu)$, $y = y(\mu)$ through (x^0, y^0) (that is, with $x(1) = x^0$ and $y(1) = y^0$) is determined by the following system of equations:

$$(A_j^T y - c_j)x_j = \mu(A_j^T y^0 - c_j)x_j^0 \qquad (j = 1,\ldots,n),$$
$$Ax = b.$$

By differentiation, at $\mu = 1$ we have the following system (where (x^0, y^0) was replaced by (x, y) for simplicity):

$$(A_j^T y - c_j)\dot{x}_j + x_j A_j^T \dot{y} = (A_j^T y - c_j)x_j \qquad (j = 1,\ldots,n),$$
$$A\dot{x} = 0.$$

Denote, as usual,

$$v_j = A_j^T y - c_j \qquad (j = 1,\ldots,n).$$

Thus,

$$v_j\dot{x}_j + x_j A_j^T \dot{y} = v_j x_j \qquad (j = 1,\ldots,n),$$
$$A\dot{x} = 0.$$

In matrix notation,

$$\begin{pmatrix} D_v & D_x A^T \\ A & 0 \end{pmatrix} \begin{bmatrix} \dot{x} \\ \dot{y} \end{bmatrix} = \begin{bmatrix} D_v D_x e \\ 0 \end{bmatrix}.$$

Equivalently,

$$\begin{pmatrix} D_v D_x^{-1} & A^T \\ A & 0 \end{pmatrix} \begin{bmatrix} \dot{x} \\ \dot{y} \end{bmatrix} = \begin{bmatrix} v \\ 0 \end{bmatrix}.$$

Let us substitute ξ_j for $\dot{x}_j$,

$$\xi_j = \sqrt{\frac{v_j}{x_j}}\,\dot{x}_j,$$

and write $D_x^{1/2} = \operatorname{Diag}(x_1^{1/2},\ldots,x_n^{1/2})$. Thus,

$$\begin{pmatrix} D_v^{1/2} D_x^{-1/2} & A^T \\ A D_v^{-1/2} D_x^{1/2} & 0 \end{pmatrix} \begin{bmatrix} \xi \\ \dot{y} \end{bmatrix} = \begin{bmatrix} v \\ 0 \end{bmatrix}.$$

Finally, this is equivalent to

$$\begin{pmatrix} I & D_v^{-1/2} D_x^{1/2} A^T \\ A D_v^{-1/2} D_x^{1/2} & 0 \end{pmatrix} \begin{bmatrix} \xi \\ \dot{y} \end{bmatrix} = \begin{bmatrix} D_v^{1/2} D_x^{1/2} e \\ 0 \end{bmatrix}.$$

It turns out that the vector $\xi = (\xi_1,\ldots,\xi_n)^T$ is the orthogonal projection of the vector $D_v^{1/2} D_x^{1/2} e$ on the null space of the matrix $A D_v^{-1/2} D_x^{1/2}$. Thus, the interpretation of the direction in terms of x is as follows. Given a pair of primal and dual interior feasible solutions (x, y), the problem (P) is equivalent to the following problem, where z is the optimization variable and x and y are fixed:

$$(\mathrm{P}') \qquad \begin{aligned} &\text{Minimize } (A^T y - c)^T z \\ &\text{subject to } Az = 0, \\ &\qquad x - z \geq 0. \end{aligned}$$

The gradient of the objective function is the vector

$$v = A^T y - c.$$

However, the algorithm takes a gradient step only after the following linear transformation has been applied:

$$T(z) = T_{x,y}(z) = D_v^{-1/2} D_x^{-1/2} z.$$

This transformation takes the current point x to the vector of geometric means of the values of complementary variables:

$$x' = T(x) = D_v^{1/2} D_x^{1/2} e,$$

that is,

$$x_j' = \sqrt{v_j x_j} \qquad (j = 1, \ldots, n).$$

The variable z is transformed into

$$\xi = T(z).$$

We thus have an equivalent problem

$$\begin{aligned} &\text{Minimize } (D_v^{1/2} D_x^{1/2} e)^T \xi \\ &\text{subject to } A D_v^{-1/2} D_x^{1/2} \xi = 0, \\ &\qquad x' - \xi \geq 0. \end{aligned}$$

Here the gradient is the same vector x' of geometric means. The projection of the gradient on the subspace of the feasible directions is as explained above.

§5. Differential Properties of the Solution Paths

In this section we consider higher-order derivatives associated with the curves $x = x(\mu)$ and $y = y(\mu)$ of primal and dual interior feasible solutions discussed in the preceding sections. For convenience, we introduce notation that is usually used in the context of the linear complementarity problem. We denote by $z = z(\mu)$ the $(n + m)$-vector obtained by concatenating $x(\mu)$ and $y(\mu)$, and we also use $s = s(\mu)$ to denote the $(n + m)$-vector obtained by concatenating the slack vectors $v(\mu)$ and $u(\mu)$. Let M denote the matrix

$$M = \begin{pmatrix} 0 & -A^T \\ A & 0 \end{pmatrix}$$

and let q denote the $(n+m)$-vector obtained by concatenating the vectors c and $-b$. Let $\dot{z}_i$ denote the derivative of z (as a function of μ), and $\dot{z} = (\dot{z}_1, \ldots, \dot{z}_{n+m})$. We also extend the arithmetic operations to vectors (applying them component by component) so, for example,

$$\frac{\dot{z}}{z^2} = \left(\frac{\dot{z}_1}{z_1^2}, \ldots, \frac{\dot{z}_{n+m}}{z_{n+m}^2}\right).$$

With the new notation the combined system of primal and dual constraints is the following:

$$s + Mz = -q,$$

$$z, s \geq 0.$$

A pair of optimal solutions is characterized by the complementary slackness conditions

$$s_i z_i = 0.$$

A solution path through a point (z^0, s^0) is determined by the equation

$$\mu\left(\frac{z^0 s^0}{z}\right) + Mz = -q,$$

which we wish to solve for z as μ approaches 0. Let

$$F(z; \mu) = \mu\left(\frac{z^0 s^0}{z}\right) + Mz + q$$

so we wish to solve $F(z; \mu) = 0$.

We can evaluate the derivatives $d^k z/d\mu^k$ by differentiating F. Let $w_i = z_i^0 s_i^0$ and let w denote the vector consisting of the w_i's. First,

$$\frac{dF}{d\mu} = -\mu w \frac{\dot{z}}{z^2} + M\dot{z} + w\frac{1}{z}.$$

Whenever a is a vector, let

$$\Delta(a) = D_a$$

denote as above a diagonal matrix whose diagonal entries are the components of a. It follows that the value of $\dot{z}$ at μ can be obtained by solving the following system of linear equations (where $\dot{z}$ is the unknown, assuming z is known):

$$\left[-\mu\Delta\left(\frac{w}{z^2}\right) + M\right]\dot{z} = -\frac{w}{z}.$$

Second,

$$\frac{d^2F}{d\mu^2} = 2\mu w \frac{\dot{z}^2}{z^3} - \mu w \frac{\ddot{z}}{z^2} + M\ddot{z} - w\frac{\dot{z}}{z^2} - w\frac{\dot{z}}{z^2}.$$

Thus, the value of $\ddot{z}$ is obtained by solving the following system of linear equations (assuming μ, z, and $\dot{z}$ are known):

$$\left[-\mu\Delta\left(\frac{w}{z^2}\right) + M\right]\ddot{z} = 2w\frac{\dot{z}}{z^2} - 2\mu w\frac{\dot{z}^2}{z^3}.$$

Notice that the coefficient matrix

$$Q = -\mu\Delta\left(\frac{w}{z^2}\right) + M$$

is the same in the equations defining $\dot{z}$ and $\ddot{z}$. It can be shown that for every k, the value of the kth derivative $z^{(k)}$ can be obtained by solving a linear system, where the coefficient matrix is yet the same matrix Q, and the right-hand-side vector is a polynomial in terms of $1/z$ and the derivatives $\dot{z}, \ddot{z}, \ldots, z^{(k-1)}$.

The fact that all the derivatives of z can be evaluated as solutions of linear systems, with the same coefficient matrix, is due to the particular structure of the function F, namely

$$F(z; \mu) = \mu\alpha(z) + \beta(z),$$

where α and β are any C^∞ maps of R^{n+m} into itself. It follows that

$$\frac{dF}{d\mu} = \mu\frac{D\alpha}{Dz}\dot{z} + \frac{D\beta}{Dz}\dot{z} + \alpha(z),$$

so $\dot{z}$ is obtained from the following system:

$$\left[\mu\frac{D\alpha}{Dz} + \frac{D\beta}{Dz}\right]\dot{z} = -\alpha(z).$$

The second derivative has the form

$$\frac{d^2F}{d\mu^2} = \left[\mu\frac{D\alpha}{Dz} + \frac{D\beta}{Dz}\right]\ddot{z} + \mu\alpha_1(z, \dot{z}) + \beta_1(z, \dot{z}),$$

and it can be proved by induction on k that the kth derivative has the form

$$\frac{d^kF}{d\mu^k} = \left[\mu\frac{D\alpha}{Dz} + \frac{D\beta}{Dz}\right]z^{(k)} + \mu\alpha_{k-1}(z, \dot{z}, \ldots, z^{(k-1)}) + \beta_{k-1}(z, \dot{z}, \ldots, z^{(k-1)}).$$

§6. Behavior Near Vertices

It is convenient to consider in this section the linear programming problem in standard form, that is,

$$\begin{aligned} &\text{Maximize } c^T x \\ &\text{subject to } Ax = b, \\ &\qquad\qquad x \geq 0, \end{aligned}$$

where $A \in R^{m \times n}$ $(m \leq n)$, $x, c \in R^n$, and $b \in R^m$. Let B denote the square matrix of order m, consisting of the first m columns of A. We assume B is nonsingular and $B^{-1}b > 0$. In other words, B is a nondegenerate feasible basis. Let N denote the matrix of order $m \times (n - m)$ consisting of the last $n - m$ columns of A.

We denote the restriction of any n-vector v to the first m coordinates by v_B and its restriction to the last $n - m$ coordinates by v_N. Thus, the objects c_B, c_N, x_B, and x_N are defined with respect to the vectors c and x. We denote by $D = D(x)$ a diagonal matrix (of order n) whose diagonal entries are the components of the vector x. Also, D_B and D_N are diagonal matrices of orders m and $n - m$, respectively, corresponding to the vectors x_B and x_N.

We assume that both the primal and dual problems have feasible regions of full dimension. The path is defined whenever a pair of interior feasible solutions for the primal and dual problems is given. Thus, let $x^0 \in R^n$ be such that $Ax^0 = b$ and $x^0 > 0$ and let $y^0 \in R^m$ be such that $A^T y^0 \geq c$. The path starting at (x^0, y^0) is given by the equations

$$x_j(A_j^T y - c_j) = \mu x_j^0(A_j^T y^0 - c_j) \qquad (j = 1, \ldots, n),$$
$$Ax = b.$$

It is obvious that for any point on this path, if we "restart" the path according to this definition then nothing changes since the products of complementary variables remain in the same proportions. Let

$$w_j = x_j^0(A_j^T y^0 - c_j).$$

Note that

$$Bx_B + Nx_N = b$$

so

$$x_B = B^{-1}(b - Nx_N).$$

Also, along the path

$$B^T y - c_B = \mu\left(\frac{w_1}{x_1}, \ldots, \frac{w_m}{x_m}\right)^T.$$

It is convenient to denote the vector in the right-hand side of the latter by (w_B/x_B). We now have

$$y = B^{-T}\left[\mu\left(\frac{w_B}{x_B}\right) + c_B\right].$$

On the other hand, for every j,

$$x_j = \mu \frac{w_j}{A_j^T y - c_j}.$$

Thus,

$$x_j = \mu \frac{w_j}{A_j^T B^{-T}[\mu(w_B/x_B) + c_B] - c_j} = \mu \frac{w_j}{-\tilde{c}_j + A_j^T B^{-T} \mu(w_B/x_B)}.$$

Suppose $B^{-1}b > 0$ is the unique primal optimal solution and $B^{-1}c_B$ is the unique dual optimal solution, so the paths of the x_j's and the y_i's converge to these points, respectively. Asymptotically, as μ tends to zero, the "nonbasic" variables, that is, x_j, $j = m + 1, \ldots, n$, are

$$x_j \sim \mu \frac{w_j}{A_j^T B^{-T} c_B - c_j} \qquad (j = m + 1, \ldots, n).$$

The denominator in the right-hand is sometimes called the reduced cost with respect to the basis B, that is,

$$\tilde{c}_j = c_j - A_j^T B^{-T} c_B.$$

So,

$$x_j \sim -\mu \frac{w_j}{\tilde{c}_j} \qquad (j = m + 1, \ldots, n).$$

Note that if y^0 is close to the dual optimal solution $B^{-T}c_B$ then we have

$$x_j \sim \mu x_j^0 \qquad (j = m + 1, \ldots, n).$$

In other words, if we start close enough to an optimal solution, the path takes us approximately in a straight line to the optimal solution. This is different from the linear rescaling algorithm where all paths tend to a single direction of approach to the optimal solution [16].

§7. Extensions to the Linear Complementarity Problem

The trajectories described in the preceding sections lend themselves naturally to the general linear complementarity problem (LCP). The problem is as follows. Given a matrix $M \in R^{N \times N}$ and a vector $q \in R^N$, find a $z \in R^N$ such that

$$\text{(LCP)} \qquad \begin{aligned} Mz + q &\geq 0, \\ z &\geq 0, \\ z^T(Mz + q) &= 0. \end{aligned}$$

Note that if z is a solution to (LCP) then for every i, $i = 1, \ldots, N$, the complementarity condition holds:

$$z_i(M_i z + q_i) = 0.$$

It is well known that the (LCP) provides a unifying framework for a large number of problems, including of course the linear programming problem. The generic algorithm for the (LCP) was developed by Lemke [14], generalizing the self-dual simplex method of Dantzig [4]. The book by Garcia and Zangwill

[8] describes the method and the general underlying homotopy principle. The interested reader may refer to this book for more bibliographical notes. The paths described here can also be interpreted as homotopies. We are interested in the behavior of solution paths in some special cases of the (LCP). There has been considerable research on classes of matrices M for which Lemke's algorithm solves the problem. It would be interesting to investigate corresponding classes with respect to the paths described here.

The general idea is a simple generalization of the case of linear programming. Given an *interior* point z^0, that is,

$$z_i^0,\ M_i z^0 + q_i > 0 \qquad (i = 1, \ldots, N),$$

consider the following set of equations:

$$(\mathrm{LCP}(\mu)) \qquad z_i(M_i z + q_i) = \mu z_i^0(M_i z^0 + q_i) \qquad (i = 1, \ldots, N),$$

where μ is a parameter. Starting at $\mu = 1$, we attempt to drive μ to zero while satisfying LCP(μ). If we succeed then we have solved the problem. However, in general we may generate a path that does not reach the level $\mu = 0$. It may, for example, diverge to infinity as μ approaches a certain positive limit. Moreover, unlike the case of linear programming, the value of μ does not always vary monotonically along a single path described by LCP(μ). Before addressing these issues, let us first consider the basic requirement of existence and uniqueness of a path through a given interior point. Consider the mapping

$$F(z; \mu) = \Delta(z)\Delta(Mz + q) - \mu\Delta(z^0)\Delta(Mz^0 + q),$$

where $\Delta(x) = \mathrm{Diag}(x)$. By classical theory, if the Jacobian matrix of $F(z; \mu)$ is nonsingular at $(z^0; 1)$ then a unique path exists through this point. Let $w = Mz + q$ as usual in the literature on (LCP) and notice that this w is not related to the weights introduced in the context of the linear programming problem. Let $D_z = \mathrm{Diag}(z_1, \ldots, z_N)$ and $D_w = \mathrm{Diag}(w_1, \ldots, w_N)$. It is easy to check that the derivative of $F(z)$ with respect to z is

$$J(z) = D_z M + D_w.$$

Obviously, $J(z)$ is nonsingular if and only if the matrix

$$\tilde{J}(z) = M + D_z^{-1} D_w$$

is nonsingular. In the linear programming problem the matrix $\tilde{J}(z)$ is positive definite (since M is skew-symmetric) and hence nonsingular for every interior point z. Obviously, whenever M is positive semidefinite the matrix $\tilde{J}(z)$ is positive definite at every interior z.

The (LCP) is intimately related to the quadratic programming problem. Consider first the following optimization problem:

$$\begin{aligned} &\text{Minimize } \tfrac{1}{2} z^T M z - q^T z \\ &\text{subject to } z \geq 0. \end{aligned}$$

An approximate problem is

$$\text{Minimize } \tfrac{1}{2}z^TMz - q^Tz - \mu \sum_i \ln z_i,$$

where μ is fixed. The necessary conditions for optimality of the approximate problem are

$$-\mu\frac{1}{z_i} + Mz = -q.$$

In other words,

$$z_i(M_iz + q_i) = \mu \qquad (i = 1, \ldots, N).$$

Obviously, we can also incorporate weights ω_i (as we did for the linear programming problem) so that the (LCP) path could start from any interior point and be interpreted as a weighted logarithmic barrier path. Specifically, if z^0 is an interior point then we can define

$$\omega_i = z_i^0(M_iz^0 + q_i)$$

and consider a path of optimal solutions (parametrized by μ) for the problem

$$\text{Minimize } \tfrac{1}{2}z^TMz - q^Tz - \mu \sum_i \omega_i \ln z_i.$$

The path is of course described by the following system:

$$-\mu\frac{\omega_i}{z_i} + M_iz = -q_i \qquad i = 1, \ldots, N.$$

It is interesting to write the defining differential equations:

$$\left(\frac{\omega_i}{z_i^2} + M_i\right)\dot{z} = \frac{\omega_i}{z_i},$$

from which it follows that $\dot{z} \neq 0$ along the path. We now consider some special cases:

(1) The matrix M is positive semidefinite.

Here the objective function is convex and the function including the barrier

$$F_{\mu,\omega}(z) = \tfrac{1}{2}z^TMz - q^Tz - \mu \sum_i \omega_i \ln z_i$$

is *strictly* convex. Thus, in this case there is at most one optimal solution to the approximate optimization problem, and it is characterized by the equations

$$z_i^T(M_iz + q_i) = \mu\omega_i \qquad (i = 1, \ldots, N).$$

Let $\omega = (\omega_1, \ldots, \omega_N)^T$. If there is an optimal solution for one value of μ then, because of nonsingularity of the Jacobian, the path extends. Moreover, uniqueness implies that μ varies monotonically. Let $z(\mu)$ denote the optimal solution as a function of μ and let

$$V(\mu) = \tfrac{1}{2}(z(\mu))^T(Mz(\mu) - q^T z(\mu)).$$

Obviously, $z(\mu)$ maximizes the sum $\sum_i \omega_i \ln z_i$ over the set

$$S(\mu) = \{z: \tfrac{1}{2}z^T Mz - q^T z = V(\mu)\}.$$

(2) The quadratic programming problem.

Consider the following problem

$$\text{(QP)}\qquad \begin{aligned} &\text{Minimize } \tfrac{1}{2}x^T Qx + c^T x \\ &\text{subject to } Ax \geq b, \\ &\qquad\qquad x \geq 0. \end{aligned}$$

The approximate function, using the weighted logarithmic barrier, is

$$F(x) = \tfrac{1}{2}x^T Qx + c^T x - \sum_j \omega_j \ln x_j - \sum_i \omega_i \ln(A_i x - b_i).$$

It is easy to derive optimality conditions using dual variables y_i:

$$-\mu\frac{\omega_j}{x_j} + Q_j x - A^T y = -c_j,$$

$$A_i x - \mu\frac{\omega_i}{y_i} = b_i.$$

As in the usual linear complementarity theory, we obtain a representation of the approximate quadratic programming problem as an approximate complementarity problem with the matrix

$$\begin{pmatrix} Q & -A^T \\ A & 0 \end{pmatrix}$$

and the defining equations are

$$x_j(Q_j x + c_j - A^T y) = \mu\omega_j,$$

$$y_i(A_i x - b_i) = \mu\omega_i.$$

Obviously, if Q is positive semidefinite then the matrix M is positive semidefinite and hence $\tilde{J}(z)$ is positive definite at any interior z. This implies that the paths converge to optimal solutions.

(3) Equilibrium in bimatrix games.

The formulation of the problem of finding an equilibrium point in a bimatrix game as a linear complementarity problem is well known (see [3]). Let

$$\Sigma_n = \{x \in R^n: e^T x = 1, x_j \geq 0\}.$$

Given two matrices A, $B \in R^{m \times n}$, an equilibrium point is a pair of vectors

$x^* \in \Sigma_n$ and $y^* \in \Sigma_m$ such that for every $x \in \Sigma_n$ and $y \in \Sigma_m$,

$$y^T A x^* \geq (y^*)^T A x^*,$$
$$(y^*)^T B x \geq (y^*)^T B x^*.$$

The matrices A and B are assumed without loss of generality to have *positive* entries. The equilibrium conditions are equivalent to

$$A x^* \geq [(y^*)^T A x^*] e,$$
$$B^T y^* \geq [(y^*)^T B x^*] e.$$

By changing variables, one can set the equilibrium problem as follows. If x and y solve the following linear complementarity problem:

$$Ax \geq e, \qquad B^T y \geq 0, \qquad x \geq 0, \qquad y \geq 0,$$
$$x_j(B_j^T y - 1) = y_i(A_i x - 1) = 0,$$

then the normalized vectors

$$x^* = \frac{1}{e^T x} x, \qquad y^* = \frac{1}{e^T y} y$$

constitute an equilibrium point. Thus, the linear complementarity problem arising from bimatrix games has the underlying matrix

$$M = \begin{pmatrix} O & B^T \\ A & O \end{pmatrix}$$

and $q = -e$.

The fundamental equations are the following:

$$x_j(B_j^T y - 1) = w_j \mu,$$
$$y_i(A_i^T x - 1) = w_i \mu,$$

where the w_i's and w_j's are positive and can be chosen to suit the starting point. Since the matrices A and B are positive, it is easy to start the paths. We can choose any $x^0 \in R^n$ and $y^0 \in R^m$ with sufficiently large components that

$$Ax^0 > e, \qquad B^T y^0 > 0, \qquad x^0 > 0, \qquad y^0 > 0$$

and then define

$$w_j = x_j^0(B_j^T y^0 - 1),$$
$$w_i = y_i^0(A_i^T x^0 - 1).$$

Consider the mapping $F\colon R^{n+m+1} \to R^{n+m}$ defined by

$$F_j(x, y, \mu) = x_j(B_j^T y - 1) - w_j \mu,$$
$$F_i(x, y, \mu) = y_i(A_i^T x - 1) - w_i \mu.$$

The partial derivative of F with respect to (x, y) at a point (x, y, μ) where

$F(x, y, \mu) = 0$ is the following:

$$\begin{pmatrix} \mu D_x^{-1} D_{w_j} & D_x B^T \\ D_y A & \mu D_y^{-1} D_{w_i} \end{pmatrix}.$$

We have not yet studied this matrix to draw conclusions about convergence of paths to equilibrium points.

References

[1] E. R. Barnes, A variation on Karmarkar's algorithm for solving linear programming problems, Research Report No. RC 11136, IBM T. J. Watson Research Center, Yorktown Heights, N.Y. (May 1985).

[2] D. A. Bayer and J. C. Lagarias, The nonlinear geometry of linear programming I: Affine and projective rescaling trajectories, AT&T Preprint.

[3] R. W. Cottle and G. B. Dantzig, Complementary pivot theory of mathematical programming, in *Mathematics of Decision Sciences*, G. B. Dantzig and A. F. Veinott, Jr. (eds.), American Mathematical Society, Providence, R.I., 1968, pp. 115–136.

[4] G. B. Dantzig, *Linear Programming and Extensions*, Princeton University Press, Princeton, N.J., 1963.

[5] B. C. Eaves and H. Scarf, The solution of systems of piecewise linear equations, *Math. Operations Res.* **1** (1976), 1–27.

[6] A. V. Fiacco and G. P. McCormick, *Nonlinear Programming: Sequential Unconstrained Minimization Techniques*, Wiley, New York, 1968.

[7] K. R. Frisch, The logarithmic potential method of convex programming, unpublished manuscript, University Institute of Economics, Oslo, Norway (1955).

[8] C. B. Garcia and W. I. Zangwill, *Pathways to Solutions, Fixed Points and Equilibria*, Prentice-Hall, Englewood Cliffs, N.J., 1981.

[9] P. E. Gill, W. Murray, M. A. Saunders, J. A. Tomlin, and M. H. Wright, On projected Newton barrier methods for linear programming and an equivalence to Karmarkar's projective method, Technical Report SOL 85-11, Systems Optimization Laboratory, Department of Operations Research, Stanford University, Stanford, Calif. (July 1985).

[10] H. M. Greenberg, and J. E. Kalan, Methods of feasible paths in nonlinear programming, Technical Report CP 72004, Computer Science/Operations Research Center, Southern Methodist University (February 1972).

[11] P. Huard, Resolution of mathematical programming with nonlinear constraints by the method of centers, in *Nonlinear Programming*, J. Abadie (ed.), North-Holland, Amsterdam, 1967, pp. 207–219.

[12] N. Karmarkar, A new polynomial-time algorithm for linear programming, *Combinatorica* **2** (1984), 373–395.

[13] C. E. Lemke, Bimatrix equilibrium points and mathematical programming, *Management Sci.* **11** (1965), 681–689.

[14] C. E. Lemke, On complementary pivot theory, in *Mathematics of Decision Sciences*, G. B. Dantzig and A. F. Veinott, Jr. (eds.), American Mathematical Society, Providence, R.I., 1968, pp. 95–114.

[15] O. Mangasarian, Normal solutions of linear programs, *Math. Programming Study* **22** (1984), 206–216.

[16] N. Megiddo and M. Shub, Boundary behavior of interior point algorithms for linear programming, *Math. Operations Res.* **13** (1988), to appear.

[17] K. G. Murty, A new interior variant of the gradient projection method for linear

programming, Technical Paper 85-18, Department of Industrial and Operations Engineering, University of Michigan, Ann Arbor (May 1985).

[18] J. L. Nazareth, Homotopies in linear programming, *Algorithmica* **1** (1986), 529–536.

[19] H. E. Scarf, The approximation of fixed points of continuous mappings, *SIAM. J. Appl. Math.* **15** (1967), 1328–1343.

[20] S. Smale, Talk at Mathematical Sciences Research Institute (MSRI), Berkeley, California (January 1986).

[21] R. J. Vanderbei, M. J. Meketon, and B. A. Freedman, A modification of Karmarkar's linear programming algorithm, *Algorithmica* **1** (1986), 395–409.